SCIENCE SIFTING
TOOLS FOR INNOVATION IN SCIENCE AND TECHNOLOGY

SCIENCE SIFTING
TOOLS FOR INNOVATION IN SCIENCE AND TECHNOLOGY

RODNEY R DIETERT
Cornell University, USA

JANICE DIETERT
Performance Plus Consulting, USA

 **World Scientific**

NEW JERSEY · LONDON · SINGAPORE · BEIJING · SHANGHAI · HONG KONG · TAIPEI · CHENNAI

Published by

World Scientific Publishing Co. Pte. Ltd.

5 Toh Tuck Link, Singapore 596224

USA office: 27 Warren Street, Suite 401-402, Hackensack, NJ 07601

UK office: 57 Shelton Street, Covent Garden, London WC2H 9HE

Cover art by Kip Ayers (www.kipayersillustration.com)

Library of Congress Cataloging-in-Publication Data
Dietert, Rodney R.
    Science sifting : tools for innovation in science and technology / Rodney R Dietert,
Cornell University, USA, Janice Dietert, Performance Plus Consulting, USA.
        pages cm
    Includes bibliographical references and index.
    ISBN 978-9814407212 (hardcover : alk. paper) -- ISBN 978-9814407229 (pbk. : alk. paper)
    1. Science--Vocational guidance. 2. Research--Vocational guidance. 3. Technology--Vocational
guidance. 4. Soft skills. I. Dietert, Janice. II. Title.
    Q147.D54 2013
    502.3--dc23
                                                                    2013007839

British Library Cataloguing-in-Publication Data
A catalogue record for this book is available from the British Library.

Typeset by Stallion Press
Email: enquiries@stallionpress.com

Printed in Singapore

Contents

List of Figures

List of Exercises

Foreword — A Personal Meditation on Creativity

Roald Hoffmann

So what's different about those who bring forth the new?

Could it be talent? I'm very suspicious of talent. I don't like it, because I want to be able to do what any man or woman can do. Music and mathematics then scare me, because demonstrated precocity in these fields seems to point to a role for talent. But then I am set at ease by chemistry, theatre, anthropology, veterinary science, high finance and university administration. There are no great chemists, actors or playwrights, or university presidents who are children. Children are inventive, unconstrained, a mirror for our romantic notions of the original good. But no child will solve Syria's problems.

It's not talent. Creativity comes, I think, from the innate, human drive to create. Into it enters discipline. Creativity derives as well from the fact that human beings cannot avoid communicating with others. And, not the least, it comes from the depths of our psyche.

I will try to explain, to myself as much as to you. First, it's in the nature of human beings to put new things here, at the very least to juxtapose the old, using chance and design to make a new array. We are incapable of not doing so — in consumable, ephemeral items such as beef Stroganoff, a Cornell quick stick, a soft contact lens. Or in lasting ways — Hans Bethe's reasoning out how the sun burns, Vladimir Nabokov's *Pnin*.

Even as we worry about the conventions of society hampering it, creation is unproblematic. But what gives the work of some people value, that of others not? There are cultural limitations to be sure, but I ask you to reflect on your reaction to pottery at the local craft fair vs. that in the strongest holding of the Johnson Museum, the oriental art on the top floor.

You can fool some of the people, some of the time, etc., but in the end value in creation emerges from discipline. Or to put it another way, from craftsmanship. Concentration, attention to detail, the measurement repeated, the play rehearsed, paradoxically frees you for the serendipitous response. That's where creativity lurks.

Good art, good science, good laws of society are acts of creation, accomplished with craftsmanship, with intensity, with an economy of statement. And set in a social milieu, of humans communicating. The impetus is likely biological — I think of the child, that marvelous machine designed to elicit a caring response. Valuable acts of creation are constructions that reach out to others; there is an easy to discern difference between poetry and therapy. We write for an audience:

Deep in
it's a docile crowd
most of the time, lazing
around, waiting for the train
of concentration to haul a few words
onto paper. It listens, then it stirs, the one
that speaks in many voices, to say:
these are just words, falling limp
into the untensed space they need sculpt, or:
make me understand.
They hate my compromises.
Here and there they offer up a phrase.
In their babble I hear the voices
of my teachers rise from a page or cafe. Sometimes
one speaks with an accent — I think
it's my father, it's him, the world
I have to please.
For them I leave no word unturned.
For it I sing, tone-deaf that I am,
the song that frees itself within.

Acceptance of the idea of an audience also demolishes for me the barrier between research and teaching. Instead I substitute a reaching out to a wider, ever-overlapping spectrum of audiences — myself, my research group, my scattered colleagues for whom I write a paper, my first year students, when I taught them.

The end of this poem touches inadequately on the psychological springs of creation. Our parents, our teachers, as benevolent as they might be, have ways of making us feel inadequate. Our religious systems and modern cultural icons are normative, so we can't live up to them. Our images of what a wife, husband, son or mother ought to be are shaped by some blend of Sarah of the Bible and Ingrid Bergman. There dwell heroes and monsters there, and the archetypical myths in which those figures meet in combat. Somewhere in this fertile murk is the tension of the master-apprentice relationship, so essential to creation: the surrender to authority, yes, authority, as a way to learning, to be followed, inevitably, by eventual rebellion.

Feelings of inadequacy fuel the creative impulse. They provide the goad to persist, sometimes the obsession to do so. The trick is to turn that inadequacy, an existential part of the human condition, into a spur. Most creative people I know do so. One key is to create obligations, for others we will do far more than we will do for ourselves.

The key to creativity is to act, in any way. In my dated days you would say, to "put pen to paper." The marvelous thing about human creation is that it has a way of growing beyond any intention of its author. True, sometimes it is like a golem, out of control. But an artifact, any artifact — a proposal for a bikeway, a rock garden built plant by plant, stone by stone, the choice to study an alkaloid from this sea sponge rather than another – intensifies as it takes shape. This is why so many have the feeling that the objects of their creation are greater than they are. This is also why any action is scary. But act we must.

... I want us to awake,
join the imperfect universe at peace with
the disorder that orders. For the cold
death sets in slowly, and there is time,
so much time, for the stars' light to scatter
off the eddies of chance, into our minds,
there to build ever more perfect loves,
invisible cities, our own constellations.

#

Acknowledgments

This book would not have been possible but for the efforts and encouragement of many people. We would especially like to thank Joy Quek for believing that, if we could pull off a book in the past, we could certainly do it again, and championing our project.

We are most grateful to the teachers, mentors and advisors who provided pivotal moments of guidance during our scholastic sojourn. For me (JD), I would like to thank Mr. Nothnaugle, my high school English teacher, who not only had us read *The Hobbit*, but insisted we invest ourselves in that world with a creative project about the book. To Professor Emeritus Richard McLain, both my university advisor and instructor for many courses, I thank you for instilling in me a love for language and the minute details of grammar, syntax and punctuation. To Professor Lois Einhorn, my Teaching of Oral Communication instructor, thank you for the precious feedback that taught me the difference between what my head thought was a worthy topic vs. how to tell what really ignited my passion.

I (RRD) would like to express gratitude to Mary Lundeen, Kenneth J. Torgerson, W. Dain Higdon, William B. Heed, Edward B. Lewis, Ray D. Owen and Bob G. Sanders. These teachers, advisors, and mentors were instrumental at key moments in my life in helping me set sail toward a future career.

Several authors deserve a special expression of gratitude for their literary works. Their words have impacted, inspired and informed us over the years. Many thanks to Martha Beck, J. R. R. Tolkien, Charles Dickens, Theodosius Dobzhansky, Dr. Richard Bartlett, and Benoit Mandelbrot.

Special thanks goes to Roald Hoffmann, Cornell professor and Nobel laureate in Chemistry, for providing the foreword for this book.

In addition to his major research contributions, Dr. Hoffmann's effort to bridge communication between the arts and sciences has served as a guiding light at Cornell and beyond.

We are very grateful to Kip Ayers (http://www.kipayersillustration.com), our favorite artist who is both intuitive and immensely talented. He provided the marvelous cover art and the Three Blind Mice figure.

Finally, several individuals were generous with their time and effort in facilitating the development and launching of the book's materials. They facilitated the new Cornell scientific creativity course offering, promoted the offering of creativity seminars or workshops, and/or beta-tested the exercises. We appreciate the support of Jerrie Gavalchin, Kip Ayers, Natalie Dadamio, Thomas Johnson, Kat Hauger, Dakota Rivers, Gerrit van Loon, Cindy Uhrovcik, Sachiko Funaba, Muquarrab Qureshi, Jon Cheetham, Avery August, Stephanie Specchio, Jill Lee, Jermalina Tupas, and Arthur Pridemore.

Introduction and Orientation to Science Sifting

A moment's insight is sometimes worth a life's experience. — Oliver
Wendell Holmes[1]

*Creativity is just connecting things. When you ask creative people how
they did something, they feel a little guilty because they didn't really do it,
they just saw something.* — Steve Jobs[2]

They thought I was crazy, absolutely mad. — Barbara McClintock[3]

BACKGROUND

Most textbooks and their related university courses that are designed to
prepare students for careers in research and technology emphasize: 1) the
nature and significance of research and 2) the tools and features needed to
move from point A to point B and maybe to point C in the initial stages of
a research career. Many educational resources are aimed at students in
graduate school who are well-embedded in the research/technology step
ladder. They are at the point of narrowing their research focus and becoming more specialized in their field. Such how-to information is critical.
Unless students and new researchers gain a measure of proficiency in
grant writing, project writing, hypothesis formulation and testing, seminar
preparation and lab operation, their career trek may never reach point B.
Those books and courses are invaluable to fledgling researchers.

Science Sifting is not meant to replace books that teach students the
nuts-and-bolts of how to enter the world of research. We will begin with
a brief list of tools students might want to use and discuss the support that
can be key to initial research training. However, we will not dwell long on
the details.

Instead, *Science Sifting* will present the reader with everything else you would likely never be taught about a career in research/technology. We will include the things you may only wonder about a decade or two into your career but that no one talks about. We will discuss those things that are self-discovered, self-learned and sometimes never learned. These well-kept secrets are the keys to promoting a creative, exciting and sustained trek through research-driven discovery.

HOW THE BOOK EMERGED

You might wonder what would motivate a veteran science researcher to set aside mainstream research in order to write a book, develop a university course and create workshops about tools for creativity and innovation in research. After three decades of pursuing scientific research, shifts in personal perception and long-overdue realizations about personal processes brought a new awareness to light. That and major changes in funding and research scope fed a need to be flexible and operate differently.

The Cornell-based, creativity-venture is comprised of two components. The first began with my (RRD's) collaboration with colleagues on the subject of the developmental origins of immune dysfunction and chronic diseases. In developing a series of concepts and papers around these topics, I began developing and using many of the nonlinear tools and techniques included in *Science Sifting*. This led to noteworthy outcomes such as ideas that seemed to emerge out of nowhere and manuscript drafts that got written in a matter of days, not weeks. Not surprisingly, coauthors were keen to know exactly what I was doing that was so dramatically affecting my scientific work output and results. The answer was hard to convey. In reality, I was using sleep to my advantage, brief moments of meditation, playing with toys and going West Coast Swing dancing more often.

As I got more questions from others and my creativity translated even more profoundly to my science, I wondered if this innovative approach to science could be taught. That was when I developed a Creative Innovation in Research seminar, my wife and coauthor created an application-based workshop, and we proved to ourselves we could indeed successfully convey these ideas to others. Hence, the course and this textbook were written in part to make it clear that I really have been playing, dancing, meditating and sleeping with a purpose, and that those activities are legitimate tools for enhancing creativity and innovation in science.

Still, having my colleagues and coauthors begging for an explanation was not sufficient motivation for me to shift from what I had been doing for three decades in order to write *Science Sifting*. A second major component was needed and a bit of synchronicity before I was ready to launch the idea. It involved a major life experience that felt very much like the types of things Roger von Oech[4] described in his book, *A Whack on the Side of the Head*. Of course, von Oech was urging readers to get out of their normal, all-too-logical, restrictive thinking. When all else fails to move you, he defaults to a whack on the side of the head.

In my case, I apparently took him too literally and actually experienced a "whack on the head" as part of a serious physical injury. It was far from what I had ever envisioned experiencing, and my whack did provide the final impetus to prepare *Science Sifting*. You could say I got to a different vantage point by attending the School of Hard Knocks. Apparently, von Oech is correct and even hard-headed thoughts can be purged.

Along the route between the "Hark Knocks" experience and this book project was an intermediate personal signpost in the road regarding the use of nonlinear tools. The signpost apparently had "TRUST" written all over it. Trust is central to the benefits you are likely to see. We will introduce many tools for enhanced creativity and provide many exercises during the course of this book that will enable you to access broader terrains of information. But what you do with the additional information becomes a personal decision. Will you trust this additional information if it comes to you via a non-traditional, nonlinear route? We invite you to ask yourself this question at the start of each succeeding chapter to determine where you are on the issue of trust. Here is the example of how this signpost worked and why it was needed before the concept could take flight.

I (RRD) awoke early one morning less than a month after the injury with a crystal-clear idea. I needed to write a paper on fractals. I did not know why I needed to write it. But there was little doubt this idea was going to stick in my awareness until a paper on fractals was produced. There was only one problem. I am not a mathematician nor have I ever been accused of being one. Nevertheless, there are aspects of the immune system, environmental health risks and patterns of diseases where a consideration of fractal dimensions can be beneficial. So in a short time, a paper concerning a fractals approach for safety of the immune system was produced. Yet, even after it was ready, I almost shelved it. Despite the burning clarity of the idea, the lack of any connection to anything I had ever done before in my career made this project seem completely implausible. After a 48 hour gut-check, the paper went in and was published.

I later received a personal message from one of the prominent leaders in my area of research stating that I "consistently come out with fresh ideas," and that this was sorely needed. It was astonishing given my severe doubt around the paper. But that was the point. I had to really trust the value of a seemingly illogical idea about fractals that came to me in my sleep before I would be ready to commit to the idea surrounding this book and the related teaching materials. Fractals represented a stepping stone in the middle of some very rapidly moving water. With *Science Sifting* we have now reached the other bank.

No need to worry about hard knocks. We will use much more pleasant tools for shifting perspectives including what von Oech[4] terms "soft thinking." Rather than slogging through the "hard way," we will help you arrive at equally empowering and useful vantage points of perception in a much simpler, easier manner.

The science presented in the upcoming chapters is sound. The outcomes have been exceptionally pleasing and useful. Where applicable, we will describe our first-hand experiences in using the tools. Mainly, though, we will describe how Nobel Prize-winning researchers have used these tools to make their key discoveries.

WHY *SCIENCE SIFTING*

The usual method for selecting a book title is to examine the content then select a title that actively conveys what the reader will encounter. It also helps if the title is catchy enough to draw the attention and interest of potential readers. At least, that is the tenet of conventional wisdom and how all the other books we have written or edited have been approached. That includes several diverse titles such as *Compendium of Scottish Silver,*

Scotland's Families and The Edinburgh Goldsmiths, Strategies for Protecting Your Child's Immune System, and *Immunotoxicity, Immune Dysfunction,* and *Chronic Disease.*

Of course giving these books titles was back before we knew and were using nonlinear tools to enhance creativity. When it came to *Science Sifting*, nothing about this book proceeded in a linear, logical fashion. Just like the idea for the fractal papers, I (RRD) woke up with the book title blazing in my consciousness after a particular night's sleep. It was one of the earliest seeds for the stream of ideas that followed for both the book and the parallel Cornell course. *Science Sifting* was part of the book proposal to the publisher. It had only one drawback. I could not even explain its meaning to my wife and coauthor, let alone to the publisher and colleagues. In fact, I used alternate titles for early lectures about the incorporated materials. So, we appeared to be stuck with a book title that just popped in from nowhere and seemed... illogical.

Soon, the book showed up in prerelease. And still I could not explain exactly what was "sifted," who or what was doing the "sifting," and to what end. All we knew was that this was the title to use. However, I may be in good company with this dilemma. When Barbara McClintock, a researcher you will hear much more about, tried to explain to her Cornell professor how she composed a piece of improvisational banjo music, her explanation was much the same. The music had just "showed up" for her, and it was the only way she could think about it. It needed no method to explain or justify it.

Ironically, *Science Sifting* was already more than half drafted when I (RRD) awoke with an unplanned chapter that proved to be the sheer essence of *Science Sifting*. Finally, the pieces fell into place and it all made sense. But until then, the title had been massively disconnected from the contents. It had been a "Cart Without a Horse." The unforeseen chapter on Sandman Science or sleep-fuzzy awareness became the core of the concept of *Science Sifting*.

It is also a prime example of what to expect when you go "nonlinear."

When operating as a linear scientist, there is usually a predictable series of events. However, when you delve into the nonlinear realm, the sequence of events and flow of information veers away from the linear. The sequence may run parallel to the linear, or things may occur in random order. In the realm of nonlinearity, Cart-Horse is just as useful as Horse-Cart. You take the Cart that shows up in your awareness and wait to see what Horse (context or pattern) it will fit into.

So, now I am no longer self-conscious to tell people the title of the book. When asked, "What is Science Sifting?" the answer is, "It's what happens during Sandman Science. Your conscious mind takes a rest, and the diligent prospector of your subconscious mind sifts through all the available information to deliver some real nuggets of insight and inspiration that then show up when you next awaken."

LESSONS LEARNED WHILE PREPARING THIS BOOK

When starting to write the book, it became apparent early on that much like the horse in Cart-Horse, I had the bit in my mouth and the project had taken on a life of its own. While we prepared a perfectly logical Table of Contents for the book proposal, upon beginning to write the first draft, it became readily apparent the contents were shifting slightly based on the research we were uncovering. Some of the original topics looked more substantial; others less so. Those are the types of minor adjustments that might be expected in any process, including the preparation of a science book. But then there were the striking departures from the originally-planned contents that we have come to term — Chapter Begetting. Here is how Chapter Begetting worked.

I (RRD) kept intending to draft one of the key, yet standard, early chapters in the book, and my coauthor/wife spent each day expecting the delivery of those first draft pages. Each day I sat down to write this chapter with a clear idea of how the material would flow. Yet, each day as I reached the second or third paragraph, I would add some new wording or phrase that would stick out. Four times I tried to get beyond the first page of the planned chapter only to be stopped by some new phrase or idea that I would be forced to engage. Each new phrase seemed so important that I could do nothing else but give the road-blocking phrase its own chapter. So, each day I spent writing a totally different chapter that was not in the Table of Contents or part of my day's planned project. It was as if the source paragraph in the original chapter comprised a portal for new ideas to leap through into my awareness. Those begotten, unplanned chapters represent about a third of the present book.

For an entire week, I was unable to move forward with the draft of the originally-planned chapter. Instead, four separate chapters flowed from my fingers. Of course, my coauthor was asking exactly what I was doing and why she had no pages to read. My only answer was "I wish I knew." Finally, after the fourth unplanned chapter was birthed, some force in the universe allowed me to proceed from page one to page two of the original chapter. Now those unplanned chapters are so thoroughly integrated into the other material that I would be hard pressed to tell which ones they were without a side-by-side comparison of the original vs. final Table of Contents. If the first question you ask yourself concerns trust of a single idea, then the second question might be: How willing are you to go with a flow that just shows up and keeps flowing? What initially looks like a roadblock that interferes with your work could be something entirely different.

THE EXERCISES

The practical side of *Science Sifting* is the Exercises included in the gray boxes within each chapter. They provide you with tools and opportunities to apply them. At the same time, you can calibrate when you notice changes in perception, available information, as well as access to new patterns and insights. We have tried these exercises, and many have been tested in classrooms, seminars, workshops and webinars we have held. Several exercises appear in *Science Sifting* for the first time.

Not every exercise is expected to be equally effective for all people. We expect that different people will gain different benefits from different exercises. What we encourage is that you try all of the exercises, even if they seem silly. Identify those exercises that had the greatest personal impact on you then find areas of your studies and research where you can apply the basics of the exercise. The idea is that these exercises may help you overcome research or career roadblocks. They may help you open to your full creative potential.

SPECIFIC AIMS FOR THIS BOOK

Since this is a book oriented toward scientific researchers and those interested in innovative technologies, we thought it would be appropriate to list some specific aims for what we would hope you will take away from this book. Whether you are encountering this book as a student, a workshop participant or simply a casual reader, there are goals surrounding the materials you will encounter in the subsequent chapters.

1. To provide you with something useful you did not know or had long since forgotten.
2. To encourage you to pursue non-work-related activities with delight, joy, energy, and the awareness that those activities can have added value for enhancing your work creativity.
3. To introduce at least one new tool that can aid you with enhanced creativity and innovation in the workplace.
4. To pleasantly surprise you with the range of tools that is available.
5. To encourage you to broaden the creative spaces at your place of work.

SUMMARY

This book about tools for enhancing creativity and innovation and the related course and workshops only happened because we paid attention to and trusted information that was not available to us though the traditional channels. We did not acquire this information by using the traditional routes that we were taught years ago in school. The ideas, specific chapters, and exercises showed up as we ourselves practiced the same tools (e.g., sleep, meditation, pattern jumping, play, music, embodied cognition) you will read about in the subsequent chapters. We invite you to test-drive these tools and find the mix that works best for you.

If useful, here is a beginning frame of reference you can use in the form of a question. You are working on a time-sensitive project and you hit a complete roadblock. Initial attempts to overcome the roadblock make it clear you are getting stuck even further. The best way to overcome the roadblock is to:

A) Stare harder at your blank computer screen for the next 3 hours.
B) Walk outside and see what you notice, take a nap, play a game, meditate, exercise, or pick up your favorite musical instrument. (Ok, if it is a piano don't pick it up, just sit down at it).

During the course of this book we hope to persuade you that while A) might work, B) has a higher success rate, is more likely to infuse your project with sparks of innovation, and is far more enjoyable.

Chapter 1

Duality and the Creative Scientist

The scientific method is a potentiation of common sense, exercised with a specially firm determination not to persist in error. — Sir Peter Medawar[1]

The greatest discovery a man could make would be to prove that one of these accepted laws was erroneous. — Michael Faraday[2]

Hell! There ain't no rules around here! We're trying to accomplish somp'n. — Thomas Edison (in response to a question from Mr. Rosanoff)[3]

The Yin-Yang of Training Creative Scientists

Traditional education has focused on the slow, methodical and incremental progress of a scientist's training. It is delivered in bite-sized, practiced methodologies designed to help the student make the next career step. This short term (e.g., 2–5 year) view of the career creates a standardized methodology toward predictable progress for trainees. This approach engrains the scientific method in each generation of up-and-coming researchers. But is this the path for producing creative scientists? Does it foster creativity and innovation? Does it really support the scientist across the lifetime of his or her career?

We argue that this method of instruction actually does the reverse. By focusing so intently on a standardized methodology with students, we run the risk of losing the innate spontaneity we begin with, or what the Root-Bernsteins term the "Sparks of Genius" in their book by the same title.[4] There is evidence to suggest that the traditional approach to educating scientists is in fact teaching the creativity right out of them.

What traditional education has overlooked is the duality of purpose or polarizing goals scientists must juggle. It is the Yin-Yang for nurturing the creative scientist. Illuminating the existence of this duality and helping students navigate it is one of the purposes of *Science Sifting*. The following descriptions best demonstrate this duality.

1. Well-educated scientists are experts focused in a significant disciplinary area. They are prepared to pursue rigorous scientific research methodologies with persistence and to engage in intense critical thinking. They have the wherewithal to doggedly pursue their hypotheses while simultaneously gaining increased recognition and credibility in their area of research. And they are in sync with current funding priorities.

2. Well-educated scientists exhibit flexibility and adaptability. They are prepared to move into new areas, define new research arenas, discard outdated paradigms and, if need be, swim against the predominant scientific current to forge new thinking. They take advantage of emerging opportunities even if they must retool, be recast differently, and/or significantly change their research setting. They are visionaries pursuing what will become recognized as important in the near future.

Most career researchers could see merits in both of these descriptions. But the education and training required to foster the first type does not translate well to the second. The characteristics and inclinations to be nurtured are not only different but virtually opposite. Our traditional education for scientists has been skewed more toward description #1 than toward description #2. However, the key for a successful, lifelong career is balance and discernment.

To persist in an outdated paradigm or area with no future is as deadly to a research career as it is to jump haphazardly from one area of interest to another without gaining a foothold in any area. Granted, different priorities may be associated with different stages of a research career. Tenured professors may feel better positioned to undertake high risks in their research. They may feel more supported in employing greater flexibility in their areas of interest. Yet, if young professors must focus narrowly in safer areas, some day they, too, will make up the ranks of tenured professors. Why, then, do we not train students for this dual role as a scientist?

Somehow, we expect researchers to already have the knowledge needed to navigate these shifts, in spite of the fact they have rarely been exposed to these competing dualities. They have been taught the strategies necessary to undertake description #1, while remaining ignorant of what will be required to engage description #2. Few full career scientists ever complete their careers still working in the area on which they focused for their Ph.D. thesis. Most scientists find they must shift their focus and goals many times throughout their careers. We wrote this book to help others navigate the less certain waters of description #2.

The education and training to achieve description #1 relies heavily on linear thinking. Description #2, by necessity, must push past that boundary into the nonlinear realm.

A Timeline

In taking a look at the preparation for a research career, it is useful to look back at the child who was open to unbounded possibilities for his or her life and forwards to the advanced career researcher. On the one hand, the child had the dreams and enthusiasm of an explorer eager to discover new territory. On the other hand, advanced researchers may periodically examine the myriad choice points during their career and wonder "what if." "What if I had made that choice?" "What if I had made that transition?"

At this point in the book, what we are presenting could just as easily apply to a career in manufacturing, accounting, social work or sports. However, there are certain aspects of research that make these states of childlike openness and mid-career, choice point "what ifs" more challenging than in other professions. The difference is inherent in the scientific method itself.

Observe and Notice

The scientific method is based on stating a clear hypothesis then methodically testing that hypothesis through a rigorous, prescribed process. Observation is the heart of this process and is mandatory throughout each stage. The successful researcher needs to employ two processes when gathering data.

First, the researcher needs to *notice* the original idea and its relationship to existing paradigms. Second, the researcher needs to *notice* the execution and flow of each stage of the research such as data (information) collection and the results. But most importantly, a researcher needs the ability to *notice* anomalies. Anomalies are the catalysts for major breakthroughs and quantum leaps in innovation.

The key to success and gaining useful insights is to observe and notice. This applies whether you're avidly pursuing the scientific method or working in a more nonlinear fashion. In all forms, you want to strive to be a keen observer capable of noticing anomalies. In their book, *Sparks of Genius*, the Root-Bernsteins[4] told their readers that all "knowledge begins in observation."[4] They went on to illustrate that most information is actually hidden in plain sight; many are simply ill-prepared to notice it. We plan to explain how to do this and give you the tools to prepare you to actually uncover what has been hidden in plain sight. The anomaly is there, but will you notice it?

You may wonder what the difference is between observing and noticing. In *The Physics of Miracles,* Dr. Richard Bartlett[5] suggests that observing is most often visual while noticing involves all of your senses. He urges readers to become aware of a feeling response by "noticing what you notice." Bartlett goes on to encourage readers to use "innocent perception" by regularly practicing the art of noticing whatever shows up without dismissing or judging it. This, in turn, allows us to move beyond our preset expectations and limiting cause-effect assumptions. The more you practice noticing anomalies, the more they will seem to show up for you to notice. It is a form of practiced activity.

The trick to "noticing what you notice" and "innocent perception" is to trust what you have noticed then to act from that trust. Dr. Bartlett[5] argues that we can open up more possibilities by focusing on what is different rather than on what is the same. Differences generally don't match our expectations and are markers of possible anomalies. Similarities keep us in our comfort zone making it more likely that we will dismiss the observation. Focusing on the differences requires a higher level of personal trust because you must step outside of your comfort zone. However, this stretch outside of that zone is where there is growth and discovery.

In *Innovation Generation*,[6] Dr. Roberta Ness discussed the importance of the anomaly and why that importance supersedes any expected data. In the case of Alexander Fleming, his experiment with bacteria failed due to the mold growing on the cultures. However, he remained open and noticed that the mold not only killed his experiment but the bacteria as well. The anomaly of the mold killing the bacteria turned out to be far more important than the data he expected to get from the bacteria cultures themselves and Fleming acted accordingly.

Sometimes a ruined experiment is that and nothing more. Sometimes the failure of an experiment turns up data far more valuable than the original experiment was designed to uncover. Some questions to keep in mind when considering this are:

1. How can you tell when a failed experiment is just a set back and when it's a stepping stone to a career-defining breakthrough?
2. Are you ready to trust what you have noticed?

Discernment

If blending linear and nonlinear approaches together is useful to research, you might also assume that having more information is as well. To a degree, this is true. We will devote a lot of time in *Science Sifting* to teaching you how to access and use the information many researchers might ignore. This will expose you to a greater spectrum of information, but this also comes with a caveat. More information does not necessarily mean better information. You must exercise better discernment when examining data sets or sifting through the flow of information. You need to extract the key pieces in order to fit together the complete puzzle and discover the useful underlying patterns. Malcolm Gladwell[7] calls this "thin slicing," which is the ability to quickly discern key information among a massive data set.

Sometimes you truly will be seeking further information. At other times you will be gleaning nuggets from the midst of the information at hand. You may not know how all the pieces fit together. You may lack the vantage point from which to see the information that is hidden in plain sight. It is like a diamond amid a myriad of glass shards. You have to be at the right angle with just the right lighting in order to find the gem.

In *Blink*, Maxwell Gladwell[7] gave an excellent example of how more information is not always an asset. Our favorite anecdote is the case in which the United States once conducted a massive military exercise in which one side had vastly smaller numbers of troops. The smaller group was led by an old-school officer known for his gut-instinct responses in battle. His group was pitted against a massive force that was outfitted with state-of-the-art technology designed to gather a wealth of data to enhance the decision making process. In a stunning fashion, the out-manned, low tech, old-school officer led his troops to soundly defeat the "superior" forces.

In this case, the massive influx of information the "superior" forces collected provided nothing but a false sense of security. The "superior" forces were unable to discern the necessary answers from it in order to predict the movements and tactics of the substantially smaller force against them. In contrast, the seasoned, old-school officer relied on the information his troops could glean along with keen instincts and efficient calculations in order to predict the linear actions of the "superior" force. In the end, it was no contest.

The computer-driven "superior" force was unable to calculate the unpredictable actions of the old-school officer despite the reams of data they gathered. On the other hand, the old-school officer easily calculated the highly predictable moves the "superior" force would use. In the end, the military experienced a massive shock that virtually unlimited information and cutting edge technology could be so easily overcome by a small number of troops led by an intuitive commander.

When it comes to information, we actually receive three kinds:

1. an overload of information
2. useful information
3. information usefully applied in terms of patterns

These should not be confused, and the only way to keep them straight is to notice anomalies and develop your ability to discern between them. One way to learn to discern the differences relates to how information comes to us. Some information arrives as individual bits or bits in strings. Other times information comes to us in the form of organized patterns. When we receive it that way, we may lock it away in the context of the initial pattern and not see all the applications.

The Trajectory of Traditional Education

Children at play seem full of excitement as well as being open to an expanse of possibilities. Through traditional research education, students gain competence, a superb level of expertise and effective communication skills. Yet, simultaneously, their perceptible scope of possibilities often gets narrowed. We say "perceptible" because the landscape likely has not changed since childhood. The students change instead.

Breaking out of the Conventional Mode

The career path shown in Figure 1.1 predictably narrows to produce a highly specialized expert. This is useful for gaining depth in a focused topic. However, this path does not introduce the student to the range of tools necessary to prepare for a lengthy research career. While students do achieve success following this model, they often do so in spite of, rather than because of, the traditional, educational approach.

The truth is the researcher will face shifts in time commitments, career-stage specific challenges, technological changes, different training circumstances, altered research priorities, and periods of uncertain funding. University programs have been slow to address these issues formally, but they are beginning to pay attention. A few institutions, such as Cornell University, are launching courses specifically to teach students nonlinear tools to promote unconventional thinking. Tomorrow's researchers need to veer off the traditional education path and seek additional experiences to help them develop greater flexibility of perception and broader research awareness. These tools will be most useful when roadblocks arise.

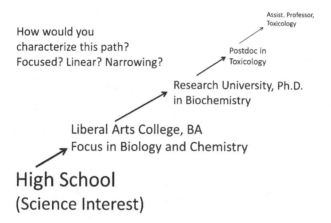

Figure 1.1   Sample Path to a Research Career

WEB TRAINING EXERCISE

We just showed one example of a stepwise education program designed to produce a Ph.D. toxicologist. However, no training program is that linear. There are always outside influences. Follow the directions below to find yours:

1.   Make a line diagram of your own training program like the earlier example from the book. This should be one you have already completed or are in the middle of pursuing.

2.   Notice where this linear trajectory is useful and also where it may be limiting.

3.   Assume this is but one long strand of a spider's web. Begin to add side connectors to this line forming a landscape rather than a straight linear path.

4.   Walk away for 5 minutes then come back and see if you notice more connectors to add.

5.   Hold onto the sheet for at least a day and continue to add to the spider's web (a grid). It doesn't matter whether these are activities or additional expertise you already have or might be able to get. If you think of it as a connector, add that strand and label the expertise, experience or activity.

6.   See the following Web of Experience example (Figure 1.2).

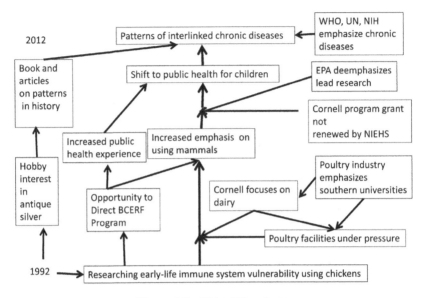

Figure 1.2   Web of Experience

Summary

In order to gain greater traction in research, it is helpful to learn tools for enhanced creativity and innovation. No single protocol is a magic elixir to maximize a given person's creative potential. Different people will identify those tools that seem to resonate best with how they learn and think. Once applied, these tools will provide you with novel perspectives on issues and greater opportunities to view your own roadblocks from different vantage points. For this reason, we do not prioritize or rank the tools we present in this book. We will try to provide you with a smorgasbord of offerings for you to sample. However, creativity for its own sake is not the goal. It is the capacity to blend rigorous, logical, scientific methodology with a healthy dose of innovative thinking, flexibility and adaptability. This can lead you to greater innovations and help you overcome career roadblocks. As von Oech[8] says, "know when to be fully logical, methodical and practical and know when to play, break the rules, cherish ambiguity and become the idea."[8] We will be stressing this merged dichotomy in the coming chapters.

Chapter 2

Moments of Scientific and Technological Discovery

I can't understand why people are frightened of new ideas.
I'm frightened of the old ones. — John Cage[1]

The history of science and technology is replete with major scientific discoveries and breakthroughs literally coming in the blink of an eye at the least expected moments. No doubt that is one reason why Malcolm Gladwell[2] chose the title, *Blink,* for one of his books. It takes but a split second for a wave of insightful intuition to hit. By paying attention to one perceptual difference, you can change a career if not the world.

Breakthroughs often occur when least expected. Many authors have described these moments of inspiration as "Ah-ha" or "Eureka" moments. Because the moments were so startling and rare, many literary reviews have analyzed every facet leading up to the breakthrough. What the reviewers have found is that scientists and inventors usually did one or more of four things:

1. They saw something such as an anomaly that others had missed.
2. They chose to pursue an anomaly that others had seen but ignored.
3. They connected the dots to identify a pattern that had been previously hidden.
4. They relinquished control of their previously stagnant career.

Most people consider moments of significant discoveries to be mysterious, unpredictable, due to scientific genius and out of an individual's control. The premise of this book, however, supports the hypothesis that while such moments of innovation are not guaranteed, they can become more likely with a broader use of nonlinear tools.

EXAMPLES OF MOMENTS OF INSPIRATION AND DISCOVERY

Mold-ing Your Better Health — Sir Alexander Fleming

Never neglect any appearance or any happening which seems to be out of the ordinary: more often than not it is a false alarm, but it may be an important truth. — Sir Alexander Fleming[3]

Scottish scientist and Nobel laureate, Alexander Fleming, made a series of discoveries that were a combination of "accidents" and keen observation of the unexpected. He spent his childhood in Lochfield Ayrshire, Scotland immersed in nature exploring the local moors, streams and valleys with his siblings. According to Alexander, they "unconsciously learned a great deal from nature."[3]

Fleming was fond of games and played them in unusual ways. In *Spark of Genius*, the Root-Bernsteins[4] describe how Fleming would play a diverse array of games for which he gave himself deliberate restrictions (e.g., playing golf with only one club or using the putter like a billiards cue while on the greens). He did this to both challenge himself and create a new experience. The Root-Bernsteins[4] noted that play is an important tool as it establishes its own rule set replacing the rules that are connected to a serious activity. By imposing his own additional rules via play, Fleming embodied a mantra we try to embrace whenever applying play to a serious issue. That is "my game, my rules."

There are some valid reasons to consider placing arbitrary constraints over your activities. According to von Oech[5] you can find new inspiration and access new patterns by constraining elements such as space, time, word number, word usage or subject. The important thing is not only to apply the constraints in your games but also the situations that emerge in real life as well.

Fleming was constantly playing and changing the rules. In fact, his mentor Sir Almroth Wright told Fleming "You treat research like a game. You find it all great fun."[4] This mode of operation in which Fleming treated his research as a great microbial game led him to more than one major discovery. In 1922, Fleming discovered the anti-microbial enzyme, lysozyme. A fellow researcher reported the Fleming noticed a strange degradation of the bacteria with which he was working. His curiosity piqued, Fleming pursued the anomaly further finding it was due to mucous secretions from a cold and went about isolating the enzyme.

In 1928, while working at St. Mary's Hospital in London, Fleming headed off for holiday to his summer home, The Dhoon in Barton Mills, Suffolk like he did every August. The home was special to Alexander because of the time he was able to spend there with his wife, son and his gardening. In fact, Maurois[3] reported that Fleming even gave up golf just to spend time with his son. Fleming was attracted to children because they, much like himself, could find happiness in simple things. Alexander loved The Dhoon, which reminded him of his childhood home back in the Scottish countryside.[3]

Before leaving for The Dhoon, Fleming cleared off a lab bench area for an assistant to use. In the process, he set aside petrie dishes he had used for growing bacteria that would have to be sorted and cleaned upon his return. The holiday provided the perfect diversion; he had time with his wife and son, while gardening and engaging in activities that hearkened back to his own childhood. When he returned to his lab on September 3, 1928, a former lab assistant, Merlin Pryce, visited. Fleming mourned the extra workload he had personally assumed since the assistant had left. To demonstrate his point, Fleming picked up a petrie dish from the discard pile and suddenly noticed the growth of penicillin mold in it and the absence of bacteria where the mold was growing.[3] The anomaly did not escape Fleming's attention, and antibiotics were born.

The Good Humor Man — Richard Feynman

Richard Feynman is one of the most widely recognized physicists of all time. He shared the 1965 Nobel Prize in Physics, played the bongos, participated in the development of the atom bomb, and was one of the more humorous professors to stroll the lecture halls of the University of Wisconsin-Madison, Cornell University and Caltech.

In Feynman's semi-autobiography[6], *Surely You're Joking, Mr. Feynman!* it is readily apparent that he was a brilliant, inspirational and, when appropriate, serious scientist. But he brought a sense of humor and play to virtually every endeavor as well. Feynman declined an offer to join the

Princeton University faculty alongside Albert Einstein and John Van Neumann in order to come to Cornell University and join Hans Bethe, his colleague from the Manhattan Project. But a few winters in Ithaca, NY sent Feynman packing for the warmth of Pasadena, CA and Caltech. He was known for scribbling on anything and everything whenever inspiration struck.

When Feynman became bored while involved with the Manhattan Project at Los Alamos, NM, he amused himself by picking the locks on colleagues' lockers and defying various secure filing systems. He discovered the lock combinations simply by willing himself to think only like a physicist (in setting a code). When this enterprise lost its glamour, Feynman would take his bongos to a local mesa and set about drumming and chanting.

In spite of the snow and cold of Ithaca, NY, it was during his time as a Cornell professor that Feynman gained his Nobel Prize-winning insight.[7] At the time, he had become disillusioned with his work and had given up making further, significant contributions to physics. He wrote: "So I got this new attitude. Now that I am burned out and I'll never accomplish anything, I've got this nice position at the university teaching classes which I rather enjoy, and just like I read *Arabian Nights* for pleasure, I am going to play with physics whenever I want to without worrying about anything whatsoever."[7] Within a week of Feynman's "letting go" of the need for further scientific accomplishments, a new insight surprised him out of the blue.

Apparently, Feynman enjoyed students so much that he joined them in the Cornell cafeteria and watched them play by tossing plates in the air. He would watch the plates spin and surreptitiously join in the students' game. Nobel Museum curator, Anders Barany, described how Feynman came to Cornell depressed about the bomb and the loss of his first wife and with waning confidence about his future research. But then one day while watching students toss plates, a "plate started wobbling in a way that caught Feynman's interest... [his observation] solved the problem he had been working on for a long time. This led him to the research that eventually resulted in his Nobel Prize."[8] Barany uses Feynman in the Nobel Museum's exhibition as an example of how playfulness can be regarded as a important type of creativity.[8]

In commemorating Feynman's breakthrough at Cornell, Susan Lang concluded that "Students at play shouldn't necessarily be discouraged. Because you never know when play might inspire a future Nobel Prize winner."[8] A replica of the actual plate that sparked Feynman's innovation is included in the permanent collection of the Nobel Museum in Stockholm, Sweden and was part of a significant touring exhibition titled: Cultures of Creativity: The Centennial Exhibition of the Nobel Prize."[8] The usefulness of play as a tool for creativity is so important that we devote an entire chapter to this topic.

The Candyman Can — Percy Lebaron Spencer

Percy Lebaron Spencer was a Raytheon Corporation scientist who had been involved in the production of the WWII era magnetron. It was developed in secret at the University of Birmingham England to help make radar detection of aerial bombing and U-boat assaults possible for the Allies. Shipped to the US surreptitiously, President Roosevelt described it as being one of the most important cargos every to reach American shores.[9] Spencer's significant design improvements led to a major production breakthrough just as the magnetron was most needed in the war effort.

The magnetron operated with a tube that used very short pulses of applied voltage. The pulses were translated into short pulses of high power microwave energy. The short-wave microwave radiation was reflected off a target, the reflection was analyzed, and a radar map was produced on a screen. Spencer's improvements allowed the magnetrons to be mounted in airplanes and smaller objects could be detected with greater accuracy.[10]

Interestingly, since Spencer was orphaned early in life, he never completed grammar school. He taught himself and gained work experience in a mill and during a stint in the Navy. With this experience, he was able to secure a job at the small, Cambridge, MA-based company, Raytheon Corporation. Here, his curiosity and inventiveness become renowned. According to Raytheon history, Spencer was allowed to take the magnetron tube home with him over weekends. He would return with it to work on Monday having made vast improvements that enhanced its operating efficiency. This led to a more than 1,000 fold increase in the daily production capacity of the magnetrons.[10]

Remarkable as this achievement was, one day Spencer was standing in front of the magnetron while testing the operation of his new tube. He had a chocolate candy bar in his pocket and noticed it had melted.[11] Rather than dismiss the anomaly, he suspected that radiation-associated heat from the magnetron had been responsible. Spencer then tested this possibility using a bag of unpopped corn. Sure enough, the operating magnetron caused the kernels to pop. He followed this with a messy experiment involving an egg.[12]

Not only noticing but acting upon his observation of the melted chocolate bar led Spencer to a patent and the first ever commercial microwave oven. The first one was named the "Radarrange" honoring the magnetron's originally intended use. It was marketed by the Tappan stove company for commercial use. Later, a subdivision of Raytheon (Amana Corporation) produced the first microwave ovens designed for home use.

Spencer became Vice President of Amana Corporation, had a building named after him, held hundreds of patents and was recognized by the US Navy with the Medal of Distinguished Service. Was this because he was a genius or because he could recognize anomalies and, once recognized, had sufficient courage and curiosity to act upon them?

Navigating the Maize — Barbara McClintock

Barbara McClintock was an American geneticist and a member of the US National Women's Hall of Fame. She took her first genetics course at Cornell University in 1921. Her performance was so stunning that she was admitted to a genetics course normally reserved only for graduate students. From there, she had an exceptional career that included a BS, MS, and PhD from Cornell in Botany as well as an instructorship. She also held professorships at the University of Missouri and the Carnegie Institution of Washington (Cold Spring Harbor Laboratories). She went on to be awarded the first McArthur Foundation grant and the Nobel Prize for Physiology or Medicine in 1983. She was awarded a commemorative US Postal stamp in 2005 under the "American Scientists" series along with Richard Feynman.

McClintock's great breakthrough was the discovery of mobile genetic elements known as transposable elements. She used South American maize with its colored kernels to make her discovery that was then verified by others over several decades.

McClintock had a childlike enthusiasm for her work as she described in her own words: "I was just so interested in what I was doing I could hardly wait to get up in the morning and get at it. One of my friends, a geneticist, said I was a child, because only children can't wait to get up in the morning to get at what they want to do."[13] She demonstrated a dogged persistence even when her scientific style deviated from that of others. She focused on finding differences rather than on looking for similarities, even if they were as small as a single kernel of maize.[14] Though her initially published discoveries were ignored by the most august scientific bodies, Barbara McClintock was driven by what has been termed her "inner knowledge."[15] This helped her to persist even in the face of rampant skepticism. Different strategies of 'informational knowing' among researchers is one factor we will discuss in later chapters where we consider the nature and location of information and how to access it.

A pivotal aspect of McClintock's discovery concerned the profound relationship she had with the plants. She was quoted as saying "I know every plant in the field. I know them intimately, and I find it a great pleasure to know them."[16] Part of getting to know her plants led McClintock to act in a type of partnership with the maize to solve specific problems. She would pose questions and watch for their responses as a tool to a very useful scientific end. It was as if the maize as her friend simply displayed its very nature to her. The answers just showed up for her. McClintock was used to things just showing up and she paid attention when they did. In fact, as an improvisational jazz banjo player, music just showed up for her as well. It was so second nature for her that she was at a loss to describe this when her Cornell music class professor once quizzed her about her approach to composition.[17]

It is ironic that in his book, *The Tangled Field*, Nathaniel Comfort[18] described how McClintock's discovery was about the search for and explanation of patterns. With a colleague, she went to a field planted with maize to view the result of a predictable experiment. But what she noticed was the pattern of the unexpected. The experimental corn field where this pattern noticing occurred is preserved today as part of the Cornell Plantations.[19] As in the case of play, we devote whole sections of this book to the concept of discovery though the processes of accessing and noticing patterns of information.

Reading, Writing and Reality — Steve Jobs

Steve Jobs, a cofounder of Apple Computers, had an unpredictable path toward innovative success. He was adopted by a family living in Mountain View, CA and was not known for his scholarship in school. He did show an interest in taking things apart and putting them back together. Though Jobs' father was a mechanic who enjoyed working on old cars[20] and Jobs observed him disassemble and reassemble things, what he was primarily drawn toward were electronics and their inner workings. As time went on, his school performance improved, and he soaked up information about electronics from other students and from engineers.[21]

Jobs spent time at Reed College in Portland taking random courses he was attracted to as a non-degree student.[22] One of these was calligraphy, the art of writing or drawing letters in different styles. Originally, calligraphy was used to decorate illuminated Medieval texts, serve as the basis for heraldic, intertwined initials used as signs of ownership on valuables, and certify royal documents, proclamations, memorials, university awards, and diplomas. So if Elliott and Simpson[23] assure us that Jobs was "bitten by the technology bug," what was he doing taking the archaic course in calligraphy? It surely seemed as if calligraphy was more a relic of an ancient past than the hope for a technology-driven future.

According to Chris Stevens, author of *Designing for the iPad: Building Applications That Sell*, the desktop computer revolution began because Steve Jobs took that calligraphy class.[24] Logic would appear to dictate that if you wanted to replace the printing press, you would learn all about printing presses. But Stevens argues that the reverse is actually true; you are more likely to be innovative if you open yourself to a horizon broader than the immediate goal. For Jobs, taking calligraphy was trusting a gut instinct and following an interest that just showed up. It was also broadening his horizon and eventually provided the route to even-spaced fonts on the Macintosh computer as part of the desktop computing revolution. In hindsight, it was precisely what Jobs would need for a critical element of his breakthrough.

Along with timely instincts for broadening his horizons, Steve Jobs seemed to have an additional, rather uncanny ability. He seemed to have the ability to shift the perceptions of his audience regarding proportions and levels of difficulty to be undertaken in a task. This phenomenon was so well documented, it was termed a Reality Distortion Field (RFD) and was included in Job's publically-released FBI files. In an interview with Amazon.com to publicize his biography of Steve Jobs, Walter Isaacson recounted the observations of Andy Hutzfeld, a close colleague of Jobs and one of the original Apple McIntosh development team members, concerning Jobs' capacity to tweak the reality space surrounding him and the effects it created. "Even if you were aware of his Reality Distortion Field, you still got caught up in it. But that is why Steve was so successful: He willfully bent reality so that you became convinced you could do the impossible, so you did."[25] Because of this ability, Jobs was able to persuade audiences as to which paths to take, and he motivated employees to accomplish tasks otherwise perceived to be impossible. We discuss the almost infectious nature of personal creativity in our chapter titled "Creative Spaces." When you carry the openness of possibilities with you as you move through the workplace, office, lecture hall, and laboratory, people simply start to react differently.

SUMMARY

Genius is not a matter of simply being brilliant, though that is certainly a plus. It is a matter of noticing and pursuing anomalies that others miss or dismiss. It is opening yourself to broader horizons and seemingly illogical possibilities and experiences. It is a matter of veering off course into less traveled territory for the sake of finding new vantage points and gaining greater perspective. Where might you find the next piece of useful information? It could be in cooking class, out golfing, picking up a guitar, bongos or a banjo or having lunch in a Cornell University cafeteria?

Chapter 3

Preparing for a Fulfilling Research Career

It is not the result of scientific research that ennobles humans and enriches their nature, but the struggle to understand while performing creative and open-minded intellectual work. — Albert Einstein[1]

The mind can see only what it is prepared to see. — Edward de Bono[2]

A career in science and technology can be one of the most satisfying pursuits you could choose. The joy of discovery can enliven each and every day. A researcher is much like an early explorer opening uncharted lands while using all of his or her faculties to find new vistas of understanding and progress. Better yet, the researcher's discoveries often translate directly into improved health, better environmental protection, comfort for those most needing it, enhanced quality of life, an adequate food supply and a deeper understanding of ourselves, our world and the universe itself. Virtually no lives go untouched by the work of researchers and technological innovators.

As with most careers, a substantial research career has its moments of pitfalls and challenges. The pitfalls can produce setbacks, and the challenges are often frustrating. However, they also provide opportunities for personal and professional growth. They present important choice points for decisions that can mold your future course.

PREPARE, YET BE PREPARED

What do we mean by "Prepare, yet be Prepared?" They sound the same, but they are not. Prepare represents the outward steps or route you take toward a goal. Be Prepared is the state describing your readiness to respond. When we talk about the duality of scientific training, the need for both action and the state of readiness is evident.

A Case Example

You need an effective plan in training for and building your research career. Important steps along the way need to be taken and can make a huge difference to the outcome. The results of these preparatory steps may not be apparent until a decade or more later. Like the developmental programming involved in later life health, these early key educational and training experiences provide the groundwork for developments later in the research career. Therefore, planning early and well is a necessity. Yet, even with the most meticulous planning, you need to remain prepared for the unexpected. Being flexible and adaptable to changing circumstances and demands is imperative. You must recognize and seize opportunities, and discern the best from among many options. Being prepared and alert is entirely different from the methodical preparation of taking coursework and conducting experiments. For this reason, we will emphasize the need to carefully plan in preparation for your research career and to remain alert and prepared to adapt to the twists and turns along the way.

A Personal Example

I (RRD) have learned a few lessons in hindsight from my nearly 50 year interest in scientific research, 35 years of which have been spent at the professorial research level. One lesson is that no matter how well-prepared you may be, things change, shifts happen, and if you are sufficiently flexible, they can be good things.

In my case, I knew I wanted to become a research scientist by 7th grade (age 12–13). The key motivator was the somewhat obscure book *The Biological Basis of Human Freedom*[3] written by plant geneticist, Theodosius Dobzhansky. Though the book was more about human nature, Dobzhansky's comments about genetics and his studies absolutely fascinated me. I wanted to know more. I declared to my parents that I was going to be a geneticist and began reading about and studying the topic at every opportunity.

As my thirst for knowledge grew, I pursued genetics training that included attending a National Science Foundation-supported summer science institute program at the University of Arizona-Tucson when I was a high school junior. There, I worked in a Drosophila genetics lab, which was also an ecological genetics lab. At the time, I was less interested in the ecology part of the program. I focused on learning about genes and inheritance patterns. I still remember being in the university student center when I watched the 1969 moonwalk.

As an undergraduate, my major was Zoology with an emphasis in genetics. I continued my laser-like focus by earning a Ph.D. in immunogenetics. I only undertook the immunology part because my lab would allow me to pursue somatic cell genetics if I also worked on hematopoietic cells. I really had no interest in blood cells or the immune system when I entered the doctoral program. At the time, it was merely a route to do cellular level genetics.

In fact, given scheduling conflicts, I never took the basic immunology course; a decade later, I headed Cornell's graduate program in immunology. I never took courses in toxicology or microbiology, either, but both have been key parts of my later activities.

So what did I become after my thesis was completed and my doctorate conferred? I woke up one morning to find myself immersed in the very antithesis of all my genetic training — the environment. When I grew up, the nature vs. nurture arguments were in full swing with little middle ground. No holistic view of how the two could fit together was considered, and epigenetics was a concept for the future. There I was fully trained and certified to perform genetics research only to be in the midst of doing environmental health science research, instead. Furthermore, I was working on the immune system, the route I had tolerated since it allowed me to do genetics research. I had virtually no paper qualifications to work in environmental health and the immune system, yet that was both what I was teaching and researching at Cornell. How did that happen? It was through a set of unpredictable circumstances and opportunities. Some doors closed while others opened.

The trend of shifts and challenges did not stop there. I went on to teach advanced immunology courses, became director of Cornell's toxicology institute, was appointed as Senior Fellow in the Cornell Center for the

Environment and have spent more than a decade teaching veterinary students about host defense against pathogens. I've done all of these without taking a basic immunology course, toxicology course, having formal environmental training or a course in microbiology. While I would have benefitted from formal coursework in these areas, the disconnect between coursework and later research activities exemplifies how varied careers can become in spite of the best laid plans and preparation.

If I needed to label my career course, I might call it "The Drunken Sailor Design" for a research career. It is not a suggested route for you to follow into your research career. However, the best possible preparation to become a geneticist is what ultimately led me to a wonderfully satisfying career as an immunotoxicologist studying environmental effects on the developing immune system. The take home messages from my example that are relevant are:

1. You do not stop learning at any point in your career.
2. Not everything you learn must come from formal courses.
3. You have significant responsibility when it comes to your own education and training, and
4. You can and should actively participate in your ongoing education and training.

FINDING YOUR FIT

One of the most universally accepted features of successful scientists as well as those in non-scientific careers is the fact that they need to enjoy, find excitement in and be sustained by their chosen career. As Steve Jobs put it, your "career is going to fill a large part of your life, and the only way to be truly satisfied is to do what you believe is great work. And the only way to do great work is to love what you do. If you haven't found it yet, keep looking. Don't settle...you'll know it when you find it....".[4]

If you are not happy with your career, your dissatisfaction will bleed over into other areas of your life. Whereas, a career you love will sustain you over the long term. In the case of research, there is a certain level of glamour. You do not know what you may discover today, tomorrow or next year. Your findings may have implications that won't be fully realized for years. From personal experience, I (RRD) have discovered that my notoriety as a researcher seems to be based on what I did 5–10 years ago rather than what I am doing today. Often when asked about those old projects, I'm not even that excited to talk about them because I would rather be talking about what I'm doing now.

There are three major steps in finding where you fit. These mesh together and don't follow a specific order or priority of importance. All three factors are foundational for building a research or technological career.

Step 1 — Find what you are passionate about.

If you figure that you will need a minimum of 7–10 years of intensive education and training before you begin making an independent and significant impact in research then your passion for that area needs to run deep.

Step 2 — Identify the subject area that both drives your passion and will still be around when you are ready to do research on it.

You can love basket weaving and can become the world's expert on basket weaving techniques above 20,000 feet, but it does not mean you will get funding to refine or improve upon those techniques.

Step 3 — Find where your natural talents reside.

You may be interested in many subjects and even passionate about several. Others may view a couple as being significant. But when choosing where to focus your energies over the long haul, ask yourself, "Which one(s) am I good at?" Always play to your strengths. You may love theoretical physics, and it is likely to continue to be important into the future. However, if calculus was a challenge and differential equations was worse, the likelihood you will become an accomplished theoretical physicist, much less enjoy the work, is slim. Leave that as a hobby interest.

Let's consider these in more depth.

Step 1 —

Having passion to bring to your career in research enables you to be rewarded in both highly tangible ways as well as less obvious ways. Much as an athlete who competes because he or she loves the game, the research experience alone can be almost as gratifying as the specific outcome. The passionate researcher awakes each morning with the knowledge that the day will be filled with something he or she loves. It does not matter whether the day's schedule appears to be seemingly trivial (e.g., waiting while something incubates) or, alternatively, a data-producing day at the end of a lengthy key experiment. There is a certain joy just in the privilege of being allowed to play the research game each day. Just as many athletes can gain enjoyment from even a well-played losing effort, the passionate scientist can be sustained even in the face of what at least, momentarily, may seem to be an experimental failure.

A key recipe for "cooking" a passion-driven career is illustrated below (Figure 3.1). What you receive during your career is directly related to what you have on hand to put into your research "stew."

Step 2 —
This concerns the area on which your research will focus and its potential for longevity. For example, an engineer in 1954–55 decided to mold his career around making major improvements in car design and manufacturing. He was determined to revolutionize the way America drove and focused on an experimental design for Ford. The project was designated the "E" car, and it arrived on September 4, 1957 to display the engineer's success. It was the first car to have warning lights for low oil or the engine overheating and had childproof door locks.[5] These were stunning technological developments for the time. But the "E" car or Edsel would only be made for two years before Ford ended production. No matter how much effort and innovation went into the car, it was a shocking failure and its collapse was complete.

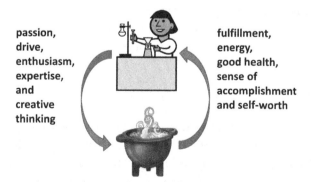

passion, drive, enthusiasm, expertise, and creative thinking

fulfillment, energy, good health, sense of accomplishment and self-worth

Figure 3.1 Thriving on Research Stew

Rarely will research priorities shift as rapidly as those around the demise of the Edsel. In the midst of the intense training and preparation for a major research focus, the challenge for the budding researcher is to gravitate toward an area of interest that not only grabs your attention but is sufficiently prominent when it comes time to apply for funding.

Step 3 —
This concerns using your natural talents. Granted, these may change over time and don't be surprised if they do. However, for planning purposes, you will need to succeed early and often in your research. It helps to pay close attention to the areas where your interests match your accomplishments. Surprisingly, students often over think this. They often pursue a trending topic of science research in the very area where they struggle most for achievement.

I (RRD) can provide a personal example of this. Early in my career, I took a Sabbatical leave to become accomplished in what were then new, cutting-edge laboratory techniques. I was planning that these techniques would drive my research in a certain direction. To put it mildly, I was a disaster in my efforts to use and import these techniques. I tried to persist in the effort, but it became clear that I would not be leading my lab forward in this direction. Twenty years later I have discovered why I was unable to master the techniques in question. I would never have guessed the explanation. I might have been able to alter my course had I known the basis of the problem, but I did not. I chose alternate routes that played to my strengths at that time. The techniques were and are still important, but they would not be for me. This can be a difficult lesson but one that is important. What is easy for some of us is not universally so for others. Find the areas where you can use your strengths.

PRACTICALITY AND UTILITY

In the last section, we discussed the need for passion about your area of research. If you are not excited about it then your work and output will lag. Others will see little reason for being enthusiastic about your research, either. Second to personal passion is the consideration of practicality and utility. If you have difficulty conveying your passion for your area of research to others, your focus may be too obscure. On the other hand, herd mentality rarely produced Nobel Prize winning achievements. Knowing when to be different, how different and what that difference is likely to mean in peer evaluation are important considerations.

If you do choose a more obscure area for research, at least one person needs to share a modicum of interest; this would be the person who will pay you to undertake the research. For greater job security, you might want to aim to identify at least two people who share your interest. What is it that captures your imagination and is an area where your participation could have tangible benefits in the future (5, 10, 20 years from now)?

EDUCATION AND TRAINING

You should try to remember that a dedicated teacher is a valuable messenger from the past, and can be an escort to your future.
— Albert Einstein.[6]

When we think about this quote from Einstein, it is easy to remember certain teachers, their mannerisms and the little factoids that we recall learning during a particular class. We may even be able to visualize the setting and the teacher's mannerisms as we were taught the factoid. But the reality is that the factoid by itself is of little importance. What did we learn beyond factoids? For example, did we learn how the factoid fits with others, and do we know where we can get the other factoids we need? Did we learn anything about ourselves, how we learn and how we interact with our world? The factoids can be a red herring. Some of the best teachers and classroom experiences involve all those learning opportunities beyond the factoids. To make this point I (RRD) will contrast my learning in the battle of Cursive vs. Ciliophora (the phylum including the protozoa, paramecium).

Perhaps in light of my universally-acknowledged poor handwriting, I can recall hours of third grade penmanship classes from a very patient and dedicated teacher. She worked so hard with me. It is a skill everyone learns, but it is also usually taught in isolation. Of course, now the hours of trying to achieve handwriting perfection are arguably of significantly-reduced importance.

In contrast, I will never forget my high school biology lab taught by a very quiet, short-of stature and very passionate biology teacher (Ms. Lundeen). Perhaps the most vivid memory involved our experience in demonstrating the regenerative capacity of the single-celled, anatomically-diverse protozoa organism, paramecium. I still remember periodically checking the cultures and my astonishment at seeing the whole organism reconstituted under particular sets of culture conditions. It was regenerative biology in action. Ms. Lundeen taught us not only factoids, but also provided us with the opportunity to better understand and fully appreciate our world. Our perceptions of how nature worked were literally hinging, in part, on what conditions allowed organisms to regenerate. Ironically, regenerative biology is one of the hottest topics in medical research today.

What Ms. Lundeen really taught me was: (1) how to be passionate about biology, (2) the joy of being astonished, (3) how to find whatever information I needed, (4) how to think critically and (5) to expect the unexpected. Whatever your discipline of interest, these basic elements of education taught by inspired teaching can escort you to your research future.

MENTORING THROUGHOUT THE CAREER

Mentoring is an aged tradition whose merits and utility have recently been revived and reapplied to researchers. In many European university traditions, a senior professor often nurtured and encouraged junior faculty until they themselves were ready to lead in that discipline. A prime example was the relationship between Almroth Wright and Alexander Fleming at St. Mary's Hospital Medical School in London, England. Wright, the head of the Inoculation Department, had distinguished himself by developing the process of antityphoid inoculations thus saving many soldiers' lives during WWI. It was Wright who persuaded the young Alexander Fleming to stay on at St. Mary's after his studies and served as Fleming's mentor. Wright's experiences and guidance certainly increased Fleming's likelihood of success. Yet, though proud of his protégé's discovery of penicillin antibiotics, Wright also warned that antibiotic resistant bacteria would eventually emerge.

In present times, mentoring has grown in academia from a once infor-mal and occasional possibility to a more universal, proactive and struc-tured effort to help young faculty. Mentoring researchers can be valuable at every level of professional development because a research career is not that different from human life stages. Some of the challenges facing a new Ph.D. are distinct from those encountered by mid-career researchers and even more different from those of a last decade career researcher. However, there are usually colleagues who have the relevant experience to guide others through those stage-dependent challenges. While we have focused on the training and guidance of Ph.D. students and post-doctoral associ-ates, there are universities and industrial sectors that have instituted men-toring for more senior researchers. They are showing that we never cease to benefit from the experience of others.

The concept of tapping more senior experience would seem to be quite simple. However, much like selecting your Ph.D. mentor and graduate committee members, finding a faculty-level mentor needs to be done with care. The process needs to facilitate open communication. Presently, uni-versities are establishing offices that have the responsibility to aid depart-ments in operating effective mentoring programs.

As part of the process, young researchers are often matched with mul-tiple mentors. A more senior researcher can often help younger faculty by reviewing grant proposals, offering strategies for publication and giving feedback on the researcher component of a tenure package. However, that person might not be the ideal individual to guide a junior professor who has questions about teaching. A better approach is to access the talents and level of experience that are needed in different areas and assign mentors based on those needs.

Ragan's and Kram's book, *The Handbook of Mentoring at Work: Theory, Research, and Practice,*[7] is an excellent reference. What becomes obvious in reading the book is that the fundamental issues surrounding mentoring programs are not confined to research or universities. They apply to a wide range of areas and cover many workplace settings. Whether it is the matching process, face-to-face vs. e-mentoring, peer-mentoring, personality fit or gender issues, the workplace is the workplace. The only real differences seem to be staffing and issues surrounding tenure in universities vs. the more fluid hiring and firing models of industry. But having annual reviews, benchmarks indicating stages of success and peer evaluations have many similarities across different professions.

Mentoring as a workplace tool has developed to a degree and has been around long enough that now various comparisons can be made. Among the resources available is the Allen and Eby[8] monograph titled *The Blackwell Handbook of Mentoring: A Multiple Perspectives Approach.* This resource compares and contrasts different mentoring models in different settings (with an emphasis on youth). Perhaps more importantly, Allen and Eby[8] also integrate concepts from these different literatures to provide practical sets of commonalities that exist across many different mentored populations and mentoring institutions.

GAINING A PERSPECTIVE

Now that we have discussed many of the elements surrounding preparation for a satisfying research career, we have an exercise for you to do to help you gain a greater perspective on your own life. We have modified the exercise from what life coach, Martha Beck, presented in *Steering by Starlight.*[9] While Beck encourages readers to take an event in the present and recognize the past events, both positive and negative, that have made the present possible, we have modified the exercise to address the past and to peer into the future.

The goal is to see the whole network of events that have made the present possible and to open your awareness of future paths that may be possible. The goal is not to lock only one path into place, but to open your awareness to options you might not have considered. Too often an event that impacts us negatively at the time elicits a response such as "Now I will never be able to...." Yet, what appears to be a dream-killer may actually be a stepping stone. It is the combination of positive and negative events that make a life-affirming path a reality. Sometimes the door to an expected path must close before the window onto an even more exciting landscape can appear.

As examples of how this can happen, I (RRD) will share two biographical sequences: my own and that of Dr. Harris Lewin, my first graduate student and a remarkably successful research scientist.

Rodney R Dietert

My quick trek from undergraduate to graduate school and a professorship seemed doomed. As a senior at Duke University, I had done two different research projects — protein biochemistry and bacteriophage genetics, been selected as one of few Rhodes Scholarship nominees from Duke, and been invited to interview for a Ph.D. program in molecular genetics at Harvard Medical School. In the 1970s, a career route in research and genetics through the Ivy League was most appealing. All went well until Harvard received my GRE scores, which were substandard by their measure. I was not accepted to the program and chose to return to my home state and the University of Texas at Austin. My chances at an Ivy League education were over. I remember thinking that, after my parents had sacrificed so much to send me out-of-state to Duke University, returning to an in-state university was a telling sign of failure.

I discovered a major consolation during a visit to UT-Austin. I would likely have the chance to work with Burke Judd, a brilliant Drosophila geneticist. Remarkably, only weeks before I moved from North Carolina to Texas, Judd committed to a position at the NIEHS and moved from Texas to North Carolina.[10] Ironically, while at the NIEHS in North Carolina, Judd held an adjunct professorship at Duke University.

Bad had just seemed to go to worse for me with Judd's departure. Yet, 3.5 years after Harvard rejected me and after I returned to my home state like a dog with its tail tucked between its legs, I was hired at an Ivy League school. Cornell University hired me as a tenure-track professor straight out of a Ph.D. In retrospect, Harvard rejecting me was the negative event that I actually needed in order to propel me into the Ivy League as a researcher. Who would have thought a rejection was actually the path forward?

Dr. Harris A Lewin

Dr. Harris Lewin is one of the most successful researchers, academicians, research institute directors and bench-to-business visionaries of our time. He was awarded the 2011 Wolf Prize, (the agricultural equivalent of a Nobel Prize for Medicine). He is the founding director of the Institute for Genomic Biology at the University of Illinois, Vice Chancellor for Research at the University of California at Davis, and a member of the US National Academy of Sciences. Among his accomplishments are: (1) the development of a pivotal strategy to protect against bovine leukemia virus, a major animal disease, (2) the pioneering of functional genomics in cattle, (3) major contributions in mapping the bovine genome, and (4) a fundamental contribution to our understanding of mammalian chromosome evolution. To honor Lewin's research accomplishments, the University of Illinois recently established "The Harris A Lewin Pioneer in Genomic Biology" Distinguished Lecture Series. Dr. Lewin's impact on animal and biomedical research has been huge.

These remarkable achievements to the betterment of animal health are very much in keeping with the contribution to research that Harris Lewin was confident he could achieve when I first met him as an undergraduate at Cornell through my immunogenetics course and undergraduate research program. Yet, there was a time when his path forward looked bleak.

In order to make a significant contribution to animal health, Harris was certain he needed a Cornell DVM degree. Yet, similar to my experience with Harvard, the Cornell Veterinary rules for entrance relied on several scoring criteria. When Lewin did not meet those standards completely, he was turned down. In fact, he was turned down twice. Instead, he completed a MS degree in my immunogenetics laboratory. He continued on earning a Ph.D. with Drs. Bernoco and Stormont in their immunogenetics program at the University of California at Davis. From there, he never looked back and is enjoying an exceptional career.

There was a moment in time when he was certain that only a DVM degree would open the doors to his dream career. However, the take home lesson from both of these biographical excerpts is that there is never just one route forward to a fulfilling research career and gaining another perspective on your life and your career may be just what you need in order to notice an alternate path and propel yourself forward.

WRITING YOUR STORY BACKWARDS AND FORWARDS EXERCISE

(adapted from Martha Beck[9])

Backwards

1. Think of a **Present Positive Event** now in your life.
2. Think of a **Positive Event** just previous that was a stepping stone to this one.
3. Think of a **Supporting Prior Event** to number 2 that occurred.
4. Think of a **Negative Event** that happened in your life without which your Present Positive Event could never have occurred.
5. Now write your story backwards using the following format:

- I was destined to......... [Present Positive event]
- Because of this destiny "x" happened [Negative event supporting the positive event]
- That allowed me to....... [Supporting Prior Event]
- This led me to........ [Prior Positive Event]
- And that helped "x" to happen..... [Present Positive Event]

Forwards

1. Think of a **Less Than Positive Event** in the present.
2. Envision a **Possible Future Positive Event** evolving from this now.
3. Envision a further **Possible Future Supporting Event** evolving from number 2.
4. Now think of a **Positive Future Achievement** that could directly follow from the Less Than Positive Event you're experiencing now through the series of Possible Positive Events.

- I am destined to......... [your Positive Future Achievement]
- This event/These events ....seemed so bad. [This Less Than Positive Event]
- To help me overcome that, this will happen...... [first Possible Future Positive Event]
- That leads me to..... [second Possible Future Supporting Event]
- And that helped me achieve..... [your Positive Future Achievement]

The point of these exercises is to show you that even if events appear dream-shattering and life-stopping, the negative events can actually be steps that propel you to a satisfying and successful future. These events are part of the interwoven fabric of your lifelong research career path.

SUMMARY

A fulfilling research career is likely to have many twist and turns. Even if you have successfully identified your strengths, found your fit, and embarked upon a path that puts those strengths to good use, you are likely to encounter many unanticipated events. These events can shape the way forward. As Martha Beck urges and we focus on in our exercise, it is best to view these events not solely as good vs. bad but, instead, as stepping stones that form the path toward where you should be. Along the way, it is important to learn from others via mentoring opportunities, and blend their experiences with your own vision and instincts. Most of all, remember that you are more than likely to arrive at your career destination via a route your never envisioned. Passion, enthusiasm and a willingness to let your career unfold are among your greatest assets.

Chapter 4
Informational Patterns

What we call chaos is only patterns we haven't recognized.
— Chuck Palaniuk[1]

Your thoughts construct patterns like scaffolding in your mind. You are really etching chemical patterns. In most cases, people get stuck in those patterns, just like grooves in a record, and they never get out of them.
— Steve Jobs[2]

Our over-arching goal for this book is to provide readers with useful tools to better recognize patterns of information. Before getting into the tools to help with informational access and utilization, it might be useful to address what information is and what patterns of information look like. Much of the content of *Science Sifting* focuses on how we perceive patterns of information, how we store and use those patterns, and how we can apply tools in order to get additional perspectives. But what exactly are we trying to perceive?

In *Meaningful Information: The Bridge Between Biology, Brain and Behavior*,[3] Anthony Reading says that "information is not an object like the neurons or semiconductors that convey it, but it is a function of the way things are arranged."[3] In other words, the pattern is integral to the information, and information is the pattern. But how can we visualize information?

We tend to visualize patterns of information as spider webs or grids. They exist parallel to each other and as bisecting webs extending across space and time. Some exist solely within the body, others are completely external to the body and still others connect parts of our internal bodies to the external information. At any one time, we can only see part of a subset of these spider webs. The goal of applying the upcoming tools is to be able to perceive more subsets of the spider webs by positioning our attention at vantage points. We like to view these vantage points as different awareness centers inside the body (e.g., the heart area) as well as beyond the body. Some of the vantage points are microscopically enhanced while others are as panoramic as an astronaut's view from space. Even your perception of the same pattern is likely to be different based on the vantage point from which it is viewed.

DESCRIPTORS OF INFORMATION PATTERNS

For the purpose of *Science Sifting*, we define patterns of information as any collection of data bits that: 1) can be found assembled together in nature, or 2) we can assemble during the importation, storage, access or utilization of information. Patterns of information are remarkably diverse, and it is nearly impossible to describe or define them. However, it is useful to describe the major features found in patterns of information.

Patterns of information can be:

1. Multi-sensory
   Patterns of information often have data pieces that are connected to our sense of smell, taste, sound, images or our general physicality. For example, if you have ever had a cold where your sense of smell was greatly reduced, you know that your food does not taste the same. You are lacking access to the olfactory part of the pattern for curry chicken, Mexican burrito, or pizza. You can still eat the food but the information you receive in doing so is incomplete. Your normal access is limited.

2. Nonlinear
   For example, you may instantly think of your mother or grandmother when you smell chocolate chip cookies baking. Another smell may remind you of when a truck hit a skunk on the road outside your house. These are fairly logical connections between smell and memory.
   However, some activities may bring up patterns of information without any seeming logic behind them. For example, whenever I (RRD) hand-wash dishes, it automatically takes me back to a neurotoxicology

conference in Arkansas in 1996 or, alternatively, to a visit with museum curators at Colonial Williamsburg about 22 years ago. The pattern connected to these events is clear and very consistently connected to the tactile sensation of hand-washing dishes. But the how and why it is connected is a matter of some conjecture. Part of the task of perceiving larger pieces of informational patterns is to follow willingly the nonlinearity of multisensory connections. Should I want to get more information about the events in Arkansas or Virginia, I merely need to wash some dishes.

3. Quite large
   One example would be the patterns of information connected to violence and war. Neither is new to civilization; each seems to have sprung up right alongside it. Given their representation in daily media, you probably have to work hard to avoid these patterns of information. With these very large patterns of information, encountering them is not the issue. The challenge for us is how to avoid interacting with them when it is undesirable.

4. Span time
   Many patterns of information cannot be restricted to one moment in time. Often, one informational pattern can be found to connect present-day data bits with bits from our childhood. If you doubt this, simply eat the food that your mother prepared most often for you (whether you presently like it or not) and see if you can do so without thinking about some aspect of your mother. It is exceedingly difficult. The food is not just the food. It is a pattern connected to data bits about your mother. When you access data about the food, you invariably get some bits about your mother along with it. Henriques[4] found that some informational patterns (e.g., prototypical shapes of human faces) remain constant both over a lifetime and even across multiple generations. In contrast, social norms shift quickly, even within an individual's lifetime. Just compare what was socially acceptable 30–40 years ago vs. today.

5. Held in multiple locations in the body
   The commonly held belief is that patterns of information are only housed, accessed and stored in the brain and cognitive structures.

However, there is a body of evidence suggesting that the entire body is used as an antenna for accessing these patterns. Massage therapists applying various types of massage that alter the neuromuscular-skeletal status are familiar with the release of strong emotions their clients often experience. Riggs[5] and Upledger[6] have both described how the massage process can open up patterns of information for clients. The emotions released represent one facet of a pattern of information. Few people realize that a back rub or a neck massage can allow for the release of deep anger or depression. The key to accessing a pattern is often via a physical change in a part of the body other than the brain. We discuss this further and provide exercises in our chapter on embodied cognition.

6. Communal
   If you are part of an identified religious community, you are not the only one engaging certain information patterns. The same holds true for cohesive cultural patterns held within strong communities such as certain ethnic groups (e.g., Japanese or Asian Indian) and indigenous cultures. These patterns often have individual, interpersonal, communal and historical connections. They can provide a well-engrained context for how one may interact with new information. This is important to understand for the researcher who is seeking a balance of dogged determination along with an adaptable view of the cosmos. Depending upon the circumstances, communal patterns can be a great friend or a significant impediment to research innovation.

   In order to understand how to look at the patterns of information we encounter, it is worthwhile to consider the more common forms in which patterns appear. Since numerous books already delve into the topic, we will only touch upon a few basics. Certain features of patterns allow us to recognize categories they may fall under but do little to illuminate their content. Nevertheless, knowing the broader forms gives us a starting point when encountering and grappling with patterns of information.

CONTENT, CONTEXT, AND CONNECTIVITY

Patterns of information we encounter may come in any shape or size and comprise a myriad of structures. As previously mentioned they are also likely to include components of information, some of which are exceedingly logical while other parts seemingly have no linear relationship to other information within the same pattern. This is perhaps the most difficult aspect of interacting with patterns. We expect they will make sense and can feel quite unsettled when the patterns veer off carrying us into uncharted territory. Later in the book we will discuss the biases and filters we bring to our interactions with patterns. But for now it is useful to highlight: 1) the multiple routes we use to connect patterns and 2) the personal nature of our interactions with patterns of information.

An example illustrating these two features can be found in how we, the authors, experience seeing operas. Recently, we attended several of the Metropolitan Opera's Live in HD telecasts at a local movie theater. Many of our senses were involved as we witnessed these multimedia comedies and dramas telecast from the Lincoln Center. (Unlike the NY theater experience, the movie theater broadcast included olfactory stimulation from the odor of popcorn. Wagner meets Orvil Redenbacher as it were.) The telecast itself had streaming audio and video information. The cues we received were both visual and auditory.

We heard the orchestra play, the singers' voices, backstage noises during interviews with the principles, audience noise from the opera house itself as well as when the cameras were rolling to capture the audience in Times Square. A richness of auditory cues was available. But that was far from the entire experience.

We saw the sets, the lighting and effects, the actions and gestures of the cast onstage, the video feed of the orchestra as they played, and the audiences from the multiple locations. Plus, we could read the visual subtitles of the French, German or Italian text translated into English. The visual cues included the form of a projected title from a highly-persuasive libretto. But it also was signaled in the singers' body language. For example, the case where the soprano's father was literally forcing his daughter to sign a loveless marriage document to further his own ambitions. In one split second, we were immersed in a multi-sensory buffet of informational cues that were woven around an operatic story.

However, the same set of cues may mean very different things to different people in the viewing audience, or even in the same household. A soprano's specific gesture or facial expression may also remind someone of a prior emotionally-connected experience and bring up memories or feelings in response. What is noble to one may be implausibly ridiculous to another. A compelling scene for one can be disturbing for another. Our receipt of cues from an informational pattern can simultaneously have both generic yet highly individualistic responses.

We have experienced this phenomenon. Among the Met broadcasts we have attended one of us will find a scene to be highly amusing while the other will find the exact same scene to be highly disturbing. One of us usually enjoys the rich musical experience and would be just as happy to listen to a recording or watch the live performance sans subtitles. The other finds the music without the actors boring and requires subtitles in order to enjoy the immersion into the story itself. Our vantage points are similar since we are sitting side-by-side, but our prior experiences and how we use information are very different. The reality for everyone in general is that patterns of information are viewed in the context of our prior experiences.

PATTERN ORGANIZATION: FRACTALS AND POWER LAWS

If you are wondering how it might be useful to think about your own interactions with patterns of information, a great place to start is with the work of Benoit Mandelbrot. As an IBM and later Yale University mathematician, Mandelbrot sought to apply mathematics for the benefit of society. His greatest legacy concerned what he termed "fractals," which are a family of geometric shapes. Fractals have the property of self-similar scaling in which there is similarity in the degree of their irregularity of shape regardless of the scale.[7] This, in turn, leads to a mathematical calculation called the fractal dimension in which the scaling characteristics can be used to quantify an object's geometry. It provides information on the complexity of a pattern as the scale is changed.

Mandelbrot found that fractals are more the rule than the exception in nature. They seem to be the basis of how complex patterns of information are organized into diverse structures. You can find fractals embedded in snowflakes, plants (e.g., broccoli), forests, coastlines, the holes in Swiss cheese, the distribution of galaxies, and our own physiological systems. Remarkably, the neurological,[8] cardiovascular,[9] respiratory,[10] and immune[11]

photograph by Janice Dietert

Figure 4.1   Fractals — Matryoshka Nesting Dolls

systems all operate under fractal dimensions, and it has been suggested that fractals can be useful in evaluating health vs. disease.[12] One example of fractal scaling is illustrated in the nested set of Matryoshka dolls shown in Figure 4.1.

The extensive occurrence of fractals in nature means that when you encounter information, it is likely to have certain geometric behavioral properties such as fractal dimensions and self-similar scaling. If you are aware of this fundamental property, you can take advantage of these scaling properties to know more than you think you know. In fact you can perform what we call 'zooming out' and 'zooming in' by using the fractal nature of a pattern of information. One example of this capacity to zoom in or out given only the information at hand is found in the study by West and colleagues. These researchers demonstrated that by using fractals, the basic structure of an entire forest is predictable based on the branching structure of a single tree in the middle of the forest.[13] That is the use of self-similar scaling in action. It is always worthwhile to question whether you can use the parts of a pattern of information that you do perceive to learn more about the boundaries of the pattern in the macrocosm and microcosm. Figure 4.2 provides a visual example for zooming in and out along a "chain" of scaling.

Figure 4.2   Zooming In and Out with Fractal-Based Patterns

In addition to using the scaling aspect of fractals to your advantage when it comes to information, there is another mathematical quirk of nature that is well worth noting. That is the prevalence of certain power laws. It turns out that hierarchies of lists of seemingly dissimilar things can have the same mathematical relationships. One of the important discoverers of the universality of power laws was George Zipf. Zipf was a Harvard University professor and linguist who was interested in word usage among the major languages. But out of Zipf's comparisons of word importance in natural languages came a fundamental understanding of human behavior and ecology. In his book, *Human Behavior and the Principle of Least Effort: An Introduction to Human Ecology*[14] originally published in 1949, Zipf posited that humans tend to communicate efficiently with the least amount of effort. He found all natural languages had the same power relationships for word usage. Remarkably, the same mathematical relationship that Zipf discovered at the root of most languages (now known as Zipf's law) extends to such unrelated hierarchies as possible opening chess moves and the rank size of cities in the US.[15] Other types of power laws exist beyond Zipf's. Awareness of these existing relationships can be useful when you are following threads of information. In fact, much of human and animal behavior follows some form of power law relationship. It is just a matter of being prepared to look for the structure.

These relationships are sufficiently important that they are being applied in the classroom to even broader issues such as predictions of human behavior. Cornell professors David Easely and Jon Kleinberg teach the implications of what might be called 'network theory' in which they examine the predictability of economics based on social networks, communication grids, crowds, games and markets.[16] The take home-message is when you first encounter informational patterns, recognize that they are part of some broader and predictable mathematical relationship. Your information predicts additional information. Find the relationship, and you can unlock a treasure trove of understanding.

SUMMARY

Patterns of information are the form through which we perceive and interact with our reality. As such, it becomes important to understand how this information is organized and how that organization can affect our capacities to both perform and excel in research. Informational patterns are often large and complex. Parts of them we may perceive as being highly logical and linearly related; in contrast, other parts may seem almost incomprehensible to us and nonlinear in their relationships. Many of these patterns have a super-organizational structure based on the mathematical relationships of fractal dimensions and power laws. The more we understand and appreciate that these relationships are a predominate theme of nature, the more we are likely to benefit from our interactions with these patterns of information.

Chapter 5

Focus on Creativity and Innovation

A scientist in his laboratory is not only a technician: he is also a child placed before natural phenomena which impress him like a fairy tale. — Marie Curie[1]

Before we present the tools that support innovation, we need to discuss what constitutes creativity. We will draw heavily from a novel public symposium, "Creativity Examined," that was held at Cornell University on July 30, 2012. Organized by Drs. John Parker, Associate Professor, James A Baker Institute for Animal Health, and Douglas McGregor, Professor Emeritus of the Baker Institute, the symposium brought together a panel of distinguished science researchers and humanities academicians to discuss creativity. The panel consisted of:

— Dr. Richard A Cerione, Symposium Chair, Professor of Pharmacology in the Department of Molecular Medicine, and Goldwin Smith Professor in the Department of Chemistry and Chemical Biology.
— Dr. Roald Hoffmann, Emeritus Frank H. T. Rhodes Professor of Humane Letters and the John A Neumann Professor of Physical Sciences in the Department of Chemistry and Biology. Awarded the (shared) Nobel Prize in Chemistry in 1981, and the National Medal of Science by President Reagan.
— Dr. Michael I Kotlikoff, The Austin A Hooly Dean of Veterinary Medicine, founding Director of the Cornell Core Transgenic Mouse Facility and Chair of the Board of Scientific Councilors of the National Heart, Lung and Blood Institute of the National Institutes of Health.

— Dr. Don M Randel, President of the Andrew W Mellon Foundation and Given Foundation Professorship of Musicology at Cornell University.

— Dr. Elizabeth Simpson, Professor and Senior Research Investigator in the Department of Medicine, Imperial College, University of London.

— Dr. Oliver Smithies, Professor of Pathology, University of North Carolina at Chapel Hill. Awarded (jointly) the Nobel Prize for Physiology or Medicine in 2007.

We will discuss the three questions addressed by this panel while drawing from the notes[2] of this symposium. The questions directed to the panelists were:

1. What is creativity?
2. What facilitates creativity?
3. What impedes creativity?

WHAT IS CREATIVITY?

The panel members seemed to find common ground when defining creativity. According to Donald Randel, creativity is "recognizing a pattern, the story in the stuff we are dealing with." Taking that a step further, Oliver Smithies defined creativity as "making a connection between things, which nobody previously has made." Elizabeth Simmons described it as having four components that included "having an open mind, looking for the patterns, breaking rules, and setting aside preconceived notions." For Michael Kotlikoff, creativity is like "opening a window in another area." He envisioned being in a totally sealed room and finding that window you could open. Roald Hoffmann described inspired creativity as an "openness to experience" aided by the "art of courage to put pen to paper."

What all of these descriptions have in common is that you must perceive patterns unseen by others and/or open vantage points onto uncharted areas. Creativity is rarely accomplished while staying within the same rule or mind-set that created the roadblock. For us, the idea of perceiving and acting on new patterns of information is fundamental to our idea of creativity and will be a major theme in *Science Sifting*. You look at the same set of information others have seen but in such a new way that you discover something fresh and new. The tools and experiences we present in this book are designed to expand your awareness of patterns of information.

WHAT FACILITATES CREATIVITY?

The panelists listed five key points as to what they felt facilitates creativity.

1. *You start exactly where you need to be as a creative individual.*

Donald Randel best described the starting point for creativity by saying that "babies are born with the openness for creativity and imagination and the challenge is to keep that alive." Infants are just beginning to establish what Roberta Ness terms "frames" and what we will view as "folders." These are the internal structures through which the infant organizes his or her world. What happens to these folders as you grow, learn more and become socialized to your culture and the world is critical. The interpersonal interactions that are necessary to produce a well-socialized adult can challenge us with more limited perceptions.

2. *You were taught to think creatively.*

Oliver Smithies described a modestly-sized English town near his birthplace that produced two Nobel laureates 20 years apart. What did they have in common? They each had the same high school teacher. This teacher chose to support their enthusiasm for learning and how to think creatively rather than to teach rote facts. In fact, most of the panel members could name teachers who had influenced them early on.

Even later-life teachers and mentors were remembered for their impact. Elizabeth Simmons described her pivotal training as an immunologist with Nobel laureate, Sir Peter Medawar. She remembered him as a polymath who derived as much excitement from his work as he did from his other interests in opera, philosophy, literature and cricket. In fact, he was purported to be able to sing his favorite operas from beginning to end.[3] In his eulogy of Medawar, David Pyke, the Registrar of the Royal College of Physicians of London, wrote: "His passion was music, as someone said, science was what he did between operas."[4] The capacity to move easily between personal interests and research is a major theme in *Science Sifting*.

While guiding students' developing lab skills as they became immunology researchers, Medawar simultaneously drew them into a broader sphere of interests. He filled his science with the same intensity of excitement that he derived from his personal interests. Simmons credited the combination of intensely focused, detailed instruction powered by Medawar's rich, non-science world as the keys to what made her a successful researcher.

Those who teach you: 1) how to access and retain your openness, 2) how to think creatively, and 3) how to engage patterns of information whatever the source facilitate your creative innovation more than those instructors who merely drill facts into you.

3. *You pursue areas that excite you and bring you joy.*

The panel members stressed the importance of loving your work and being excited about it every day. A research career is not something to survive; it should be a path to success that makes you thrive. When you enjoy your work, you may be more aware of spaces for creativity. Oliver Smithies advised student researchers to "pick something you enjoy doing. You will not be creative if you don't enjoy it. Every day enjoy." Your passion and excitement for your research can remind you of the openness and wonderment you experienced as a child.

It is very difficult to be truly creative and innovative if you hate what you are doing. However, we would like to distinguish between loving what you do and loving your job. You may love what you do passionately, yet intensely dislike certain aspects of your job setting. These are work-related patterns. Job settings can shift and some of the tools in *Science Sifting* can help make that possible. But if deep down you do not love your career path, your childlike openness slams shut. Finding the personal and career-related activities that thrill you and bring you joy is the key to bringing innovation to the forefront.

4. *You are not locked into a narrow path forward.*

Michael Kotlikoff emphasized maintaining the flexibility to shift your career. He started university life as a literature major and described how this has aided him in his present science career. He shifted out of literature and into biomedical genetics and physiology. While his literature experience was important, it could not predict the twists his career would take.

Richard Cerione also emphasized maintaining flexibility of mind and suggested it was integral in having a career progress toward greater creativity and fulfillment. Again, Sir Peter Medawar is probably the penultimate example of maintaining flexibility of mind as his interests rested on many topics. Besides science, he published books such as: *The Uniqueness of the Individual*[5], *Aristotle to Zoos: A Philosophical Dictionary of Biology*[6], and his autobiography, *Memoir of a Thinking Radish.*[7]

To encourage mental flexibility, Roald Hoffmann suggested putting yourself on the opposite side of your own hypothesis. Test out that different mindset and see how that perspective informs your original concept. He also suggested working in a community where no idea is dismissed as being crazy. He stressed that "out of considering that no idea is crazy, you can come to some important ideas yourself."

5. *You find the tools that aid your creativity.*

The panel identified several tools that they deemed most beneficial in aiding creativity.

A. Sleep

Oliver Smithies declared that sleep is one of his best tools for producing creative thoughts. He works on a problem right before bedtime, goes to sleep, and often awakens about 3 am with a possible solution. He keeps writing material near his bed, jots down the idea in the dark (to avoid waking his wife) then falls back to sleep. Invariably, his early morning insights propel him forward in his work.

Sleep is so well recognized as benefitting creativity that we have devoted an entire chapter to it.

B. Storytelling

Two of the panel members stressed the role of storytelling in creativity. Donald Randel urged the audience to find "the story in the stuff we are dealing with." And Roald Hoffmann agreed that "storytelling is impor-tant." Both he and Oliver Smithies indicated that conversations with others can promote storytelling.

C. Teaching others

Teaching is related to storytelling. In the process of examining a concept to convey to others, two important opportunities emerge. One, you may be able to develop the concept into a story. Plus, you have the opportunity to place yourself in the position of the students you will teach. When you examine your original concept from their perception, you gain a different vantage point. In fact, Roald Hoffmann urged the audience to "put yourself in others people's minds" because doing so frees your own constraints around concepts and ideas.

Oliver Smithies assured the audience that "teaching is not just a chore; it can be a great help for creativity." Donald Randel noted that people who enter a discipline less oriented to their prior training often ask the most challenging questions. Teaching others is an excellent venue for freeing your own mind.

D. Lateral thinking

Lateral thinking is similar to the concept of pattern jumping that we will introduce later. Lateral thinking and pattern jumping involve applying concepts derived from progress in one area directly to an unrelated issue in another area. In the process of lateral thinking, you may apply the nature of one pattern to help enlighten a different problem. Or you can apply the progress made in discerning one pattern to a completely unre-lated pattern elsewhere.

Dr. Richard Cerione introduced the idea of lateral thinking by referring to one of Medawar's great discoveries that was described in *The Eureka Moment: 100 Key Scientific Discoveries of the 20th Century.*[8] In this example, Medawar was working with a burn patient. He noticed that a series of skin grafts donated by the patient's brother grew for about 15 days. The second set of grafts applied after the 15 days failed to take. In thinking about the problem, he recognized the pattern of timing was similar to the two weeks it takes for an individual's immune system to respond to a vaccination. In recognizing the pattern similarity between the skin grafts and vaccination responses, Medawar developed the hypothesis that transplant rejection was an immunological phenomena.[8]

WHAT IMPEDES CREATIVITY?

While not a main focus during the symposium, interesting perspectives about what impedes creativity arose during the course of the discussion. If flexible thinking aids creativity, then inflexible thinking impedes it. For instance, Elizabeth Simmons discussed the utility of having a broad range of interests. She pointed out that having passionate interests beyond his research was a key to Medawar's creativity. In contrast, one can surmise that isolating oneself in the lab does not provide the richness of experience to enhance creativity. Similarly, Oliver Smithies advocated engaging in rich communication rather than isolation to stimulate novel ideas. He cited opportunities that occur during retreats and even at intellectually-rich monasteries. However, not all forms of communication are equally helpful in promoting creativity.

For one, formal education can be constraining. Donald Randel emphasized that while all children are born creative, we tend "to beat it out of them." He identified society and media as "organized to constrain our perceptions." This point raises a red flag on education and the search for a better way to nurture the inherently-creative child.

Elements of modern life such as news headlines, politics, economics and grant funding can act as wet blankets on the fires of creativity. The daily ebb and flow of drama and emotion that trigger fight-or-flight thinking modes can freeze our brains into significantly narrower vantage points. You become like a horse with blinders on running down the racetrack and unable to see anything but the narrow path in front of you. And researchers can trace the effects of this stressful way of operating in our bodies. We discuss this in our chapter on embodied cognition.

If you doubt how much formal education has impacted thinking flexibility and how much modern culture has subtly built boxes around you, try the following exercise.

THE AIRPLANE EXERCISE [adapted from von Oech[9]].

1. Get 60 sheets of paper, preferably those already destined for recycling. Divide them into 3 stacks of 20 and place on a table near you.
2. Mark off a line in the room between 10–12 feet away.
3. Using the first stack of 20 sheets of paper, make paper airplanes. See how many you can make fly across the line in 90 seconds. Record how many made it for trial #1.
4. Repeat step 3 using the second stack of sheets, again with a 90 second time limit. Feel free to vary how you make your paper airplanes. How many flew across the line this time? Record how many made it for trial #2.

Prior experience can certainly help while performing the Airplane Exercise. On the second trial, most people are more attuned to making paper airplanes and may have gotten better at releasing the airplanes, so a small increase in success is likely. However, it is usually just a small increase because you're more efficient with the mechanics.

By the way, how did you make your paper airplanes? What guidelines did you follow? Where did you get your information for what a paper airplane looked like? Did you use your prior experience with paper airplanes? Did you follow prior instructions or suggestions from others?

OK. Now for trial #3.

5. With the final stack of paper, crumple the sheets into balls and see how many you can throw across the line in 90 seconds. How many airplanes flew across the line? Record the number for trial #3. Has the number increased from your previous two trials? If so, by how much?

Did you feel like you were cheating when you crumpled the paper into a ball and threw it? Did throwing balls of crumpled paper defy your concept of a paper airplane? Who told you what a paper airplane had to look like? Did we specify a particular model? Did the conventional model for a paper airplane serve you well in this exercise?

In an exercise like this in which you have a short amount of time and short flying distance, a crumpled ball of paper usually works better than what you're used to calling a paper airplane. This is similar to what happens when it comes to scientific creativity and innovation. When under pressure, it becomes too easy to rely on conventional models rather than finding the best model that works within the given constraints.

If time hadn't been a constraint in this exercise, you could have made 30 beautifully crafted, standard model paper planes and flown them all across the line. Likewise, if the line had been double or triple the distance, the only models that would have gotten across the lines would have been the standard paper airplane models. But those constraints were not in place. This exercise is similar to the reason architects developed skyscrapers. Space constraints for accommodating large numbers of people on tiny plots of land required an innovation in design that forced buildings to go vertical.[9]

In some cases, new concepts are so different from the standard that they are akin to opening up another dimension in your references for ideas. We will refer to these sea changes in perception as "going multidimensional."

IS THERE A USEFUL RECIPE FOR CREATIVITY AND INNOVATION?

Any recipe for creativity in science probably starts with marrying some level of expertise and persistence with a healthy measure of flexibility and adaptability. We have developed four recipes for creativity, and they have one thing in common. Above all, they start by asking you to throw away, or at least suspend, the conventional rules you learned during your formal education. After all, look what it got you — the present problem or road-block. To get around the roadblock, you need to follow a different path.

Scott Berkun stressed four key elements in planting and nurturing the seeds to innovation in *The Myths of Innovation*.[10]

1. Innovation has many parents. Virtually every type of motivation you can imagine can serve as a fulcrum and drive the energy for

innovation (e.g., sorrow, love, pride, ego, envy, service, sacrifice). Don't discount even sorrow and tragedy in your life as part of a route to innovation.

2. Make your own beginning rather than seeking some magic elixir of some other past genius' beginning. Put pen to paper, strike a few keys on the computer or do an initial browser search. See what shows up. Once you start, you have the opportunity to gain perspective and explore. Merely having started is tantamount to taking a step on a path and, as a result, being able to see more of the path ahead.

3. Actively bring expertise to the table. Don't passively be a bystander of history. This is work with direction or what we would view as "effective work." If you want to make the work hard, that's your choice. We prefer easy, yet effective.

4. Become adept at shifting directions. In my (RRD's) personal experience, I once lost a nearly completed scientific paper from my computer hard drive with no back up file as a rescue. When I tried to recapitulate the paper, it changed drastically from the first draft. It was not only a completely different paper but much better as well. The profound lesson for me was that you may need to lose or shed something in order to shift direction.

Berkun[10] stresses that, while many innovations started out headed in one direction, directional change became the key to unexpected, but useful outcomes. One example is 3M Company's Post-It Notes. They only came about because someone kept the apparently useless discovery of "weak glue" around, and someone else found a use for it.

In the second recipe for innovation, von Oech[9] emphasizes the need to:

1. Practice looking for more than one right answer.
2. Break the rules you have imposed upon yourself.
3. Use logic and practicality when helpful; otherwise set them aside.
4. Become comfortable with ambiguities; shades of gray are a beautiful thing.
5. Imagine how others would approach the problem.
6. Play frequently.

7. Don't place a value judgment on apparent success or failure. You don't know where they are leading you if you maintain a short term view only. For von Oech,[9] failure can be portals through which new opportunities arise.
8. If need be, get whacked upside the head now and again to shift yourself out of conventional thinking.

In *Innovation Generation*,[11] Roberta Ness emphasized using tools to break out of the constraints of our prior experiences and see things differently when needed. She has a number of exercises that are aimed at undoing the conventional frames that bind and limit thinking. One powerful tool Ness employs is to change the questions you ask. She also encourages setting yourself in opposition to your favorite hypothesis in order to force a broader perspective. Only when you have identified and incubated innovative ideas is it time to show the dogged persistence necessary to carry you past others' criticism.

Lastly, Robert and Michele Root-Bernstein[12] offer several suggestions to enhance creative success including:

1. Rethink thinking. Don't focus on *what* to think; focus on *how* to think.
2. Learn to use the images that pop into your awareness.
3. Recognize patterns.
4. Use analogies and metaphors. Practice finding the relatedness among otherwise dissimilar objects.
5. Employ body-thinking. Use what some term your sixth-sense or whole-body experience.
6. Think multidimensionally. Move from using 2-D to 3-D space in your thinking.
7. Empathize. See the world through the eyes of others.
8. Use models to enhance imaginative skills
9. Play. Make up games within the realm of science.
10. Transform. Use tools for creativity to generate transformative interactions particularly among members of a team.
11. Synthesize. Integrate the tools you use into a multimodal, unified framework.

SUMMARY

We have highlighted the key points from the landmark Cornell University symposium on creativity as well as some of our favorite authors' views on the tools that are useful to break old, ingrained and limiting ways of thinking. Many of the Nobel laureates, exceptional researchers, panel moderators, leaders of the humanities, and creativity authors identified the same useful tools. Not surprisingly, we will home in on those tools and add others taken from a diversity of sources and our own personal experiences. Now that we have explored what a fulfilling research career might look like, we will dive into the pool of discovery with a toolbox full of tools and a toybox full of ideas and exercises designed to help you: (1) trust yourself more with the information you have available, (2) get the most of what you already know (but perhaps don't know that you know), (3) get started, (4) move at will to different vantage points, (5) identify, follow and jump between patterns of information, (6) make your hobby the most effective work you have ever done, and (7) rediscover and learn from your inner child.

Chapter 6

Mind Your Language

The words seemed to bite physically into Gatsby. — F. Scott Fitzgerald[1]

If you grew up in certain parts of the world during the early-to-mid 20th century, your mother probably told you to "mind your language." This was meant to warn you not to swear or use the popular phrases "everybody" was saying but instead to use proper, polite speech instead. We also find it useful to "mind your language," though we have a different intent in mind. We suggest that you are far better off watching the words and phrases you use, eliminating those that keep you stuck in unhelpful patterns and paradigms and seeking new wording that will, instead, increase your capacity for creativity and innovation.

When we use language, we not only communicate our thoughts and feelings to others, we also communicate and shape our own reality. Through your words, you tell a story with yourself as the central character. You instruct yourself as to where you have been, are presently, and plan to go. Frequently, you do all of this unconsciously without purposeful intent. Even so, your body responds neurochemically and physically to the message. The signals it receives are very clear even if the words you repeat to yourself are habitual or may be words that someone else told you in the past.

In *Change Your Words, Change Your Life: Understanding the Power of Every Word You Speak,* Joyce Meyer[2] points out that we literally eat and digest our own words. Whether you choose your words by intent or by habit, you feed yourself those words. This includes all of those words you run through your head but never voice. This internal self-talk can impact your level of joy, peace and physical energy.[2] When you consider all the words you hear yourself think and say, those messages impact your entire sphere of operation. You are narrating your life. Those words are all part of the story you tell yourself and feed yourself on a daily basis.

Language has the capacity to reinforce rules, yours or those others taught you, and in the process those rules limit your sphere of operation. But purposeful words can also free you and open you up to new possibilities. Your words can rigidly define your world as an existence of stark dichotomies: black and white, good and evil, right and wrong. Alternatively, you can choose words that describe an infinite array of potentials and possibilities. If you have never examined the specific words and phrases you use in work and life, this is the perfect time to start. Even subtle shifts in self-talk and the language you depend on daily can produce significant changes in your creative potential. If you have been practicing careful word selection, this is the opportunity to calibrate how well your newly chosen words are working for you. Find out what story your words tell and how you are defining your life.

LANGUAGE AND ROADBLOCKS

How you identify and describe your roadblocks and your ability to remove them is determined by the language you use. Whether the roadblock involves your entire career or just this week's experiment, the words you use define the playing field. If you are a scientist, engineer or mathematician, chances are you speak in exacting, precise terms. In fact, you most likely had this approach drilled into you. Plus, you will have gained experience in effective public speaking to enable you to adequately present your research and technological findings. This means you are already, or are preparing to become, a persuasive public speaker with a need to use exacting language and specific words and phrases in your presentations. However, since one of your career goals is the effective, persuasive transmission of ideas, the messages you give yourself are even more powerful. Your training in descriptive precision may well box you into a corner at times. Roberta Ness[3] emphasizes that the language you use reinforces your potential rigidity (your frames). Your self-messages can impact your self-belief.

For example, how do you describe problems you encounter at work? The word-choice of "problem" with which you use to label a work issue already creates its own box. Meyer[2] suggests a preferable way to characterize a roadblock is to call it a "situation" rather than a "problem." Situations tend to just be something to move beyond, whereas problems carry a yoke-like burden with them. Dr. Richard Bartlett[4] goes further to suggest you are better served to call what we normally define as "problems" as just "stuff." Problems define and delineate hard boundaries whereas "stuff" is more like white noise.

We know from personal experience that problems can loom as something huge and daunting, whereas stuff is often easily pushed aside. Stuff is just some fuzzy, immediate grouping of something. Stuff is much easier to rearrange into a more useful situation. You are less likely to label stuff as either good or bad. "Stuff" is the type of fuzzy language that provides you with greater wiggle room, which in turn helps you to see the path for moving beyond the roadblock.

WORDS EQUATED WITH ACTION: HISTORICAL PERSPECTIVE

Linguistic studies seem to show that the impact of spoken language was better understood in earlier times. At one point in time to say a thing was the same as doing it. Language was the equivalent of a physical act and follow through was a must. For example, in Victor Hugo's once banned 1832 play, *Le Rois'amuse*, and Verdi's opera, *Rigoletto*, that was based on the play, the court jester uses his sharp-tongued barbs to inflict pain on others while entertaining the court. Then one day the tables are turned and a horrible series of events are set into motion by a mere return of words from one of Rigoletto's unfortunate victims. The epitaph muttered by Count Monterone was understood by 19th century audiences as being equal to physically-mediated vengeance. Today, we have to be reminded why the mere utterance of a few words produced abject fear in the cowering Rigoletto. What this mindset believed and what science is beginning to prove is that the phrase we were taught as children, "sticks and stones may break my bones but words can never hurt me," is blatantly false. Words cannot only harm us, they can lock us into mundane, mediocre thinking for perpetuity.

To illustrate this further, take the Hugo play, *Hernani* (1830), which Verdi converted to the opera, *Ernani* (1844). In this story Ernani makes an oath that swears his life in exchange for the immediate right to carry out a specific deed. The pledge Ernani swears to his competitor has the power of the force of action in motion, one that must be carried out at a later date. When Ernani's life is called due by the villain, there is no recourse for him. To act against his oath would be unthinkable. Ironically, Ernani does have a choice. He could die worthy of family honor and the woman he loved, or live disgraced and unworthy of her love. The story was clear and compelling in the 19th century, yet it is almost incomprehensible to audiences in the 21st century for whom language no longer conveys this same requirement of action.

Another example is the evolution of the word "novel," since we are encouraging novel thought in *Science Sifting*. According to the Online Etymology Dictionary,[5] "novel" as an adjective arose in the 1400's meaning "new, young, fresh." The noun "novella" came into vogue in the late 1500s and meant "new stories." However, by the time of Charles Dickens in the 1800s and the development of the full novel, the noun "novel" was tightly associated with works of fiction. That means that today the same term signifying a fresh, new idea also means a fictitious narrative. It is safe to say that if this linguistic ambiguity has any effect, it may be to increase the skepticism of new ideas. Carter[6] explores this intriguing dilemma in a discussion of the cultural history of creativity.

LABELS

Words have the capacity to identify us and define our permissible activities. They are as powerful as the physical symbols like specific clothes, hand gestures and tattoos that identify membership in a specific gang or the brand that identifies specific ownership of a herd of cattle. You may even intentionally label yourself as a way of clarifying your self-identity (e.g., I'm a Cornellian) and as a form of solace. Sometimes you will have to label yourself in order to market your research expertise (e.g., I'm a geneticist) and progress within scientific circles. In fact, one of the first things that journals, publishers, granting agencies, large scientific societies and even news media do is ask you to self-select your labels. This guides them in how they will interact with you per your grant proposal, manuscripts to review, chairing scientific sessions at conferences and providing media comments on research findings. You are also going to have to label yourself in order to register with certain science-oriented groups. Just be aware that in labeling yourself, you may be assuming a certain risk when it comes to creativity.

When you label yourself or are labeled by others, intentionally or unintentionally, you can become locked into the confines of that expertise or mindset. This is what Martha Beck terms "the illusion of fixed conditions."[7] In her description of reality, nothing is truly permanent, so it is more useful to avoid speaking as if things are rigidly fixed. She actually advocates refraining from using verbs of "being" when speaking of yourself, words like: is, are, am, was, has been, will be, hasn't, weren't, isn't, aren't,[7] because these words identify your existence as something specific. For example, instead of saying "I am fat," which identifies you with your fat such that if you lost the fat you would no longer exist, it's more preferable to say "I carry extra weight." If you carry something, you can always set it down without your existence being threatened.

In my (RRD's) case, my academic title provides the label of Professor of Immunotoxicology identifying me as an Immunotoxicologist. The label has expectations and boundaries that do not include publishing in history and physics journals. Nor does the label suggest that teaching creativity to researchers and writing on the topic is within this box. There are days when I speak and write on immunotoxicology and days when I do not. This is where Beck's suggestion of carrying names, titles and labels and being able to set them aside without losing a sense of existence is more useful than embodying them 24/7. Then you can start describing things actively moving from existential phrases like "I am...." to active phrases like "I interacted with...." or "I carried...." In my (RRD's) specific case, "I research immunotoxicology" and "I explore Scottish history," plus "I talk about creativity."

Ironically, in the 1990s my (RRD's) research activities had changed so extensively that I petitioned, with the support of my department chair, to have my academic title changed. While this is possible to do, it is a relatively rare occurrence in academia and takes considerable effort. It is far easier to think of having different hats and switching them at will. Doing so, rather than acquiring labels, provides greater flexibility.

You, your relationship to patterns of information, and the names you give yourself and these patterns are critical to your overall perspective.

It can frame you in ways you may be unaware of and not recognize. For example in a 2008 seminar,[8] Dr. Bartlett referred to the word "sin" as an acronym representing the words Separate, Identify and Name. In this case, naming or labeling an information pattern becomes the basis of making a judgment and causes you to lose a desirable neutral observer state.[8]

METAPHORS

The topic of how to stimulate creative thinking has been tackled many times across the centuries. During this time, the idea of what constitutes the mind has shifted and changed. However, the one tool that returned again and again was the use of metaphors. One of the most straightforward definitions is that metaphors "are comparisons that show how two things that are not alike in most ways are similar in one important way."[9] One of the more entertaining, accessible, yet insightful books about metaphors was written by Dr. Mardy Grothe.[10] In his book, Grothe argues that the best metaphors can achieve a type of permanent residence in our minds.[10] Their transformational potential can been seen such that a "metaphor is a kind of magical mental changing room, where one thing, for a moment, becomes another, and in that moment is seen in a whole new way."[11]

The prototypical phrase used to indicate creative innovation is the metaphor "thinking outside the box."[3] Lakoff and Johnson[12] describe it as "imaginative rationality" meaning the metaphor has both an element of imagination coupled with grounded reality. Metaphors provide a way to "comprehend partially what cannot be comprehended totally."[12]

While metaphors can be beneficial to creative thinking, they can also be detrimental. The utility of metaphors seems to depend upon their origins and prior use. Some metaphors become so old and outdated, they lose their novelty and become part of everyday facts. Norquist[13] cites the phrase "body of an essay" as an example. This metaphor has become a structural label for essays rather than a freshly used, anatomical association. Stale, rigid metaphors can lock us into a box much like labeling your area of expertise can pigeonhole your opportunities and the flexibility to move into new areas. They can be pervasive, laden with emotion and habitual in an unfortunate way.[3]

An example of an unhelpful grouping of metaphors is the repeated association between time and money. This association occurs in phrases like time is money, spend your time, worth your while, use your time profitably, wasting my time, time to spare.[14] In fact, David R. Hall,[15] points out a pattern in which many of the words we use to talk about money are the same ones we use to talk about time. The same commonality can be found in talking about the movement of people and words pertaining to water as in the following description: the **flow** of people at the farmer's market **dried** up by noon.[15]

While you may have to work to get paid, our metaphors involving time-money and hard work-results help to solidify the conventional thinking that you must put even more time into your work if you want to see more results. The pervasive idea, that if you only put more time into your work you will finally succeed in life, has undoubtedly led to many burned-out, shortened careers. We work hard every day to thwart this "work-yourself-into-an-early-grave" metaphoric association. In fact, our personal mantra is more akin to: "more creativity, less work, more results."

You challenge yourself to perceive patterns of information that had been outside of your awareness when you use metaphors. Practicing their use is a lot like exercising a specific muscle at the gym. In this case you exercise the muscle responsible for perceiving patterns with the result that you enlarge your capacity to do so.

One way this idea has been applied is in education. Pascolini and Pietroni[16] developed a strategy to teach highly technical and complex physics surrounding Feynman diagrams to students without the application of high level mathematics. They took the diagrams and presented them as metaphors. What is useful in the classroom for expanded learning can be applied to research as well.

Sagarin and Gruber[17] suggest that metaphors can help you understand unfamiliar references by pairing them with what is known. By using meaningful associations, learning and creativity are enhanced. The authors stress the utility of building "ensembles" of metaphors. Begin with a single metaphor and extend it to families, fields and collections of grouped metaphors. Rather than being an interpretive act, the collection of metaphors is really a construction of webs with concepts being at some nodes and metaphors being at others. Because both the grouping and their size are personal choices, creating webs of concepts and metaphors provides for an element of improvisation.[17] This act alone will provide creative openings.

METAPHOR ENSEMBLE EXERCISE

1. The twelve metaphors below are designed to provide an array of abstract and concrete ideas.
2. Pick your favorite of the twelve to serve as your foundation and create an ensemble by adding three more metaphors to go with it. The additional metaphors should be based on one of the subject words.
3. Repeat the exercise creating an ensemble based on the author rather than the common subject.
4. Try the exercise again using another metaphor as the foundation for a new ensemble.

All the world's a stage (subjects — world or stage; author — Shakespeare)

There is more to life than simply increasing its speed (subjects — speed, time, life; author — Mahatma Ghandi)

There's a force in the universe that makes things happen; all you have to do is get in touch with it. Stop thinking ... let things happen ... and be ... the ball. (subjects — force, universe, control, allow, sports, ball, embody; author — for this, use any metaphors ever associated with any of the actors in the movie, Caddyshack; e.g., Rodney Dangerfield, Chevy Chase, Bill Murray, Ted Knight)

A computer is like a bicycle for our minds (subjects — computer, bicycle, transportation, mobility, mind, perception, awareness; author — Steve Jobs)

All religion, arts and sciences are branches of the same tree (subjects — religion, arts, science, branches, tree; author — Albert Einstein)

Life is a journey, but don't worry, you'll find a parking spot at the end (subjects — life, journey, destination, space, home, car, parking, end; author — Isaac Asimov)

Courage is grace under pressure (subjects — courage, grace, pressure, stress; author — Ernest Hemingway)

America has tossed its cap over the wall of space (subjects — hat, tossed, cap, wall, space, exploration; author — John F. Kennedy)

(Continued)

METAPHOR ENSEMBLE EXERCISE (*Continued*)

Life is not divided into semesters (Subjects — life, semesters, academic schedule, school; author — Bill Gates)

Forgiveness is the fragrance that the violet sheds on the heel that crushed it (subjects — forgiveness, fragrance, flower, foot, crushed; author — Mark Twain)

Dwell as near as possible to the channel in which your life flows (subjects — existence, living, river, flow, movement, energy, vitality; author — Henry David Thoreau)

Words are of course, the most powerful drug used by mankind (subjects — words, humans, drugs, pharmacological effects, altered behavior; author — Rudyard Kipling)

MIXED METAPHORS AND THE RESET BUTTON

The mixed metaphor appears as an error in word linkage in a phrase that brings the brain to a screeching halt. Mixed metaphors are made up of an incongruous combination of elements taken from other existing metaphors. Usually, they are multiple metaphors that have been torn apart and reassembled in ways that make no sense at all. Some examples of mixed metaphors are:

A leopard can't change his stripes.
Here is your bat so you can score a touchdown.
He is so green behind the ears.
I shot the wind out of his saddle.

What does your mind do when you read these? For us, we were thrown completely out of our normal mindset as if someone had just hit the reset button. Mixed metaphors provide the opportunity to disengage from the entanglement surrounding us. For us, reading them gave us the chance to restart our thought process and find a different path to overcoming a roadblock. You are likely to find it useful to introduce a few mixed metaphors into your most challenging days. Refresh your operating system and disengage from problematic entanglements.

SUMMARY

The words we and others select to describe ourselves, our activities, our ambitions and our surroundings have a tangible effect on our physical and sensory modes of operation. We embody these descriptions. We can select words that provide us with unbounded potential or alternatively, use words that severely narrow our options via the story we tell ourselves. Our language, including use of metaphors, is critical to the flexibility or rigidity of our informational folders. With care, we can learn to nudge these folders and enhance our creative potential aided by our language. Careful use of metaphors and the avoidance of limiting labels are useful tools for this process. Finally, mixed metaphors are the perfect jolt when you are really stuck in a rut — or perhaps we should we say barking up the wrong stone unturned.

Chapter 7

Perceiving Your Reality

Everything you see or hear or experience in any way at all is specific to you. You create a universe by perceiving it, so everything in the universe you perceive is specific to you. — Douglas Adams[1]

...the distinction between past, present and future is only a stubbornly persistent illusion. — Albert Einstein[2]

All our knowledge has its origin in our perceptions. — Leonardo da Vinci[3]

All that we see or seem is but a dream within a dream. — Edgar Allan Poe[4]

Every man's world picture is and always remains a construct of his mind and cannot be proved to have any other existence. — Erwin Schrodinger[5]

Much of this book is about how, as an existing researcher or a researcher-to-be, you can increase the likelihood for creative thinking and innovative breakthroughs over the course of your career. If a more innovative research future is where you want to travel in your career, then it becomes important to recognize exactly where you are as you begin this trek.

WHERE ARE YOU IN YOUR REALITY?

As a child, I remember local TV stations running public service announcements asking "It's 10 pm, do you know where your children are?" We would like to turn that question around. "You're ___ years old. Do you know where you are?"

The majority of people never ask themselves two simple questions: "Where am I? How did I get here?" No, not the route you took to school or work today. Although changing that around or noticing something

74

different next time you traverse it could be useful. Instead we mean, how do you perceive your present reality (i.e., your professional lot in life), and how did you get to this place? Are you where it is joyful, exciting fulfilling, novel, unexpected and adventurous? Or is your life predictable, consistent, methodical, safe, conventional and socially appropriate? Perhaps it's a mix of the two?

Answering these questions can tell you a lot about your own relationship to the reality you perceive. Why are these useful questions to ask? In part, because I (RRD) spent more than a couple of decades of my research career never asking myself these questions. Then, too, I likely avoided the questions because I didn't want to hear my own answers. As a result, I often felt frustrated when trying to find a point from which to shift my career toward a more useful route. Answering these questions may also illuminate where you are in your present comfort zone and can show you what entrenched filters you have in place that color the information you gather and use. Your comfort zone and your filters can lock you into the reality you perceive.

THE STEP #1 EXERCISE

1. Write down in 100 words or less where you are and how you got there. (Write down the first thing that pops into your head.)

From here on out, use this for a starting line and for comparisons. Are you in a joyous, exciting, if unpredictable place? Do you want to be? If you aren't, what would such a place look like?

At the heart of these questions is the tenet that your perceptions are likely to be biased and inaccurate. Thankfully, perceptions are flexible. This point is key to molding an innovative research career. Your personal history and internal narrative of how you got to where you are today are all drawn from your perceptions about your reality and yourself. These perceptions are likely to be constrained and less accurate than you think.

WHAT ARE YOU REALLY SEEING, TOUCHING, HEARING, TASTING AND SMELLING?

We will begin with a seminal paper by Dale Purves[6] of Duke University, Beau Lotto of the University of London, and their colleagues that was published in the U.S. Proceedings of the National Academy of Sciences.[6] These researchers have dedicated their careers to identifying the processes that produce a disconnection between what we think we see and what is actually there. While the focus of this PNAS article is on vision and perception, the same principles extend to information from all your other senses as well. Or as Dr. Richard Bartlett says, "Our unconscious rule-filters determine what shows up in our world and equally what is denied conscious access."[7]

THE THREE BLIND MICE RULE

When perceiving reality, we like to refer to the Three Blind Mice Rule. The Three Blind Mice are the three factors that cause us to perceive a reality that is only a crude approximation of what is actually there. While we think we are perceiving our world accurately, in reality...not really. Each of the Three Blind Mice (Figure 7.1) causes us to operate with incomplete or skewed information. To align the reality we perceive with actual reality, we

CONTEXT   HISTORICAL USE   SELECTIVE ATTENTION

artwork by Kip Ayers

Figure 7.1   Three Blind Mice

need to recognize how and when the Three Blind Mice are in operation. These Mice/factors are: Context, Historical Use, and Selective Attention.

Mouse #1: Context

Context may be the background of an image in a picture. It can be a specific odor or the expected color of a rose. Context affects how we perceive information regardless of which of the senses is involved. Because our experience generally streams information through multiple senses at the same time (the crackle of wood, smell of smoke and the sight of dancing flames from a fireplace), contextual-based errors can occur among all the senses represented. For example, would you be able to feel or generate a cold sensation if you stuck your hand into a fire? Likely not since the broader context of a fire is usually associated with heat. If a cold fire were to exist, an adult would be unlikely to sense it. In contrast, a toddler who has not yet come to equate heat with fire would be perfectly fine with feeling a cold flame.

Professor R. Beau Lotto of the University of London and Lotto Labs studies how our senses use information and make order out of that information. He has basically discovered how we construct and define our reality.

We take in information and our brain places it into a context. This context is based on our prior experiences. While we see information as patterns, if we have never encountered a pattern before, we will have difficulty perceiving it. Using context to interpret information as images is a rapid method of making sense of the world so we know how to function. However, if what we are seeing is new to us or different than our historical model, our brain will make errors. We won't see what is actually there; we will only see an approximation that will likely deviate in an unlikely way. In most cases, we will be completely unaware of these errors and anomalies.

Once you consider the fact that you largely define your reality, you may have several reactions.

1. This is utter hogwash and I wasted my money on this book.
2. Who messed me up so bad in grade school that my senses now lie to me?
3. How can I identify and reduce the errors within my current perceptions?
4. What are the possibilities that I actually define my reality?

VISUAL PERCEPTION EXERCISE

This exercise is our version of the visual exercises Beau Lotto has developed to illustrate the importance of context on our perception of reality. Look at the stars in Figure 7.2. Which of the stars is darker? In fact, they are exactly the same shade of gray. But notice how the shade of the surrounding box (the context for the stars) affects how dark the stars appear to you. Context matters.

Figure 7.2   Context Matters (Similar to Examples from Beau Lotto and Dale Purves)

Examples of the significance of context and drawing upon prior experience (for patterns) can be found in certain language exercises. What do you perceive when it comes to letters and words? Try to read the phrases in Figure 7.3.

Instructions: Read the following three phrases out loud

Yu   cn re d th s if u t y.

Ar y u re di g th s n w?

W at ar y o re di ng?

* After similar exercises by Beau Lotto and Roberta Ness.

Figure 7.3   Prior Use of Letter Strings*

Were You Able to Read the Statements?

How did you know what letters to use for any that were missing? Did you assume the words were in English and not some other language? It is likely that you made assumptions that were based on your prior experience.

Mouse #2 — Historical Use Of Information (Developmental Programming)

Dale Purves of Duke University often shows two images for comparison.[6] One is of an actual cube. The other is of a computer-generated, graded, textured and shaped surface that fools the eye. One is a 3-D item; the other is a 2-D image. But since our only historical reference is for a cube shape, we label both the 2-D and the 3-D images as cubes. Due to this, we automatically "see" shadow lines as if the flat surface we are viewing were a cube. The implications of this are quite staggering. When you begin to realize our brains deceive us, not just when viewing optical illusions but on everything we "see," you have to give yourself permission to expand your awareness of what is actually there.

If you are still skeptical how present-day perceptions are programmed by previous experience, here are two additional examples for consideration.

Consider persons with post-traumatic stress disorder (PTSD). These individuals have previously experienced highly stressful, life threatening, traumatic situations. For example, soldiers returning from an active war zone are at a higher risk for PTSD than the general population. With PTSD the brain and body become hot-wired for a fight-or-flight response that is triggered by a sensory cue. In the case of veterans, they might suddenly swerve while driving to avoid a pothole (previous cue — landmine). A car backfiring might make them duck under furniture for cover (previous cue — gunfire, grenade explosion). They might get highly defensive if others quickly encroach on their personal space (previous cue — assassin, suicide bomber). What for most people are innocuous cues may be perceived by the person with PTSD as something dangerous based on previous experience of trauma.

Another example of previously useful, developmentally programmed perceptions involves ingrained perceptions of body composition and form. The perception we hold can overwhelm even contemporary sensory input with respect to body awareness. Individuals who have lost a limb often continue to sense its presence. These perceptions can persist for years.[8] Previous experience tells the brain that a limb should still be there and sensations (e.g., pain) are elicited that go with this possibility.

Very overweight people experience a similar effect if they have been overweight for decades and suddenly lose a lot of weight. Many of these individuals will look in the mirror and still see themselves as being overweight even though the image entering their eyes is different.[9] Conversely, individuals with anorexia can look in the mirror and see themselves as fat. Yet, when shown a photograph of their body, they may recognize that it's too thin.

The visual cues simply don't compute with all the developmental programming that these individuals have acquired. According to researchers at McMaster University, perceived changes to body image are far more important than the actual changes themselves.[9]

Mouse #3 — Selective Attention

The third Blind Mouse is Selective Attention. As researchers, we are taught to focus, focus, focus in order to gain depth in our specialty. However, the process of intense focusing enlarges our blind spot for certain perceptions. In drilling down with our attention, we literally go subterranean in our perspective. Christopher Chabris and Daniel Simons demonstrated this elegantly in *The Invisible Gorilla*.[10] Presented in books, videos and demonstrations, their work shows how easy it is to miss the obvious when you narrow your focus.[11]

The basic demonstration video (http://www.theinvisiblegorilla.com/videos.html) involves teams dressed in contrasting colored uniforms (e.g., white vs. black). The viewer is usually asked to specifically focus on the white (sometimes red) team and count how many times they pass the ball between their teammates. Both teams move in circular and interwoven patterns with the dark team waving their hands to try to block the passes of the white team. If you focus as carefully as requested, you will have an accurate count of how often the ball was passed, but you will have missed when a gorilla dances through the middle of both teams and exits the other side. The gorilla is invisible to perception if you have narrowly focused. This selective attention means that in the hyper-focus of our scientific research activities, we almost certainly miss prominent anomalies. Our attention is focused exclusively on the expected range of outcomes; unexpected events are screened from view.

ANOMALIES ANYONE?

The combination of the Three Blind Mice — Context, Historical Use and Selective Attention — should be a cautionary message to researchers. These three factors mean we are at high risk for missing some of the most important observations of our research careers. Have you already done that? Are you likely to do that tomorrow? Have you visited a Cornell cafeteria lately just to observe?

What are you doing to eliminate your blind spots including your skewing or screening of what is actually there? What can you do? Surprisingly, the simplest take home answer is to recapture your childhood. While the different exercises in *Science Sifting* have varied bases and goals, the underlying connection is that they are geared to take you back to where you once had a childlike openness. Children expect to see something new every day and they do. Plus, they get excited about it.

As adults, we expect and in some cases rely on seeing sameness. In fact, deviations from sameness can be quite unsettling to some adults because it conveys an element of loss of control.

What Did You Expect?

If the adult default is to see sameness, that tendency has implications. Habituation tends to occur over time. Repeatedly seeing the same thing reduces the magnitude of signals that occurs in our visual neurons.[12] Your brain becomes anesthetized to your environment. It becomes as dull as the world around you appears to be. When novelty is introduced, neuron responses change. The more oddball the stimulus, the more likely you are to have those neurons fire. This is something we can control to an extent. Start by shifting your expectations from noticing sameness to noticing novelty. Doing this may require a certain amount of trust. If you automatically view differences as a threat, then shifting expectations to frequent differences may be challenging. Trusting yourself is a pivotal step for reducing your approximation error for reality.

Perceptional Variation

Our reality is both literally contextual and unique to us. It is not going to be the same as another person's. One way to examine personal realities is to compare notes and memories with colleagues right after an important seminar. There will be points that overlap, but you will have picked up key points that your colleagues missed and vice versa. Sometimes you may wonder if you even attended the same seminar. The reality behind the anomalies between notes and memories is that the speaker gave a lecture based on his or her perceptions. The audience heard overlapping yet different points being made. Each audience member brought their personal experiences to the lecture. These existing patterns helped them organize the lecture information in ways that made sense and were useful to them specifically. Roberta Ness refers to this method of organizing information as placing information into frames.[13] In this book, your frames are represented as your desktop folders.

Of course, we have now just provided students who fall short on exams with the perfect excuse. No longer did the dog eat their homework. They either didn't have the appropriate historical folder or lost track of the folder they needed. It could be time to back-up your "hard drive" before your next birthday.

Shared Realities

The formation of widely-accepted scientific paradigms is, in essence, a shared reality among scientific experts. In the case of a paradigm, a majority of researchers have agreed to perceive that the world operates according to a specific set of rules. The group filters information in coherence with one another and will be consistent with a particular outcome. An example is the paradigm that gravity is one of the strongest forces in the universe. Lots of people benefit by sharing in this reality. Not many have "Flying off the face of the Earth" scheduled on their day planner. The context surrounding "gravity as a strong force" is historically useful.

Yet, there are those scientists who believe gravity may actually be a weak force or something else altogether.

When it comes to risk aversion and paradigms, things get interesting. For example, many people view airplane flights as a high risk undertaking. Their view is that air travel is much more risky than driving to the store for groceries. Yet, statistics prove the opposite to be true. An auto accident near home is far more likely than being in an airplane accident. The perception of risk doesn't match the reality. So, why are so many people misled by their perceptions? One possibility is their greater familiarity with cars as opposed to air travel. Rather than "familiarity breeds contempt," in this case it is "familiarity defines the comfort zone."

This is one of the reasons we emphasize later in the book the need to shift the entire neighborhood when it comes to flexibility and the acceptability of innovative change. But if your "neighborhood" stubbornly clings to existing, shared-reality paradigms and you challenge them, good luck with that. It may be a rocky road ahead. A shared reality can be important for continuity of ideas and a basis for common understanding. However, when new information arises and sheds doubt on the conventional paradigm, it could be time for it to shift.

Ness[13] discusses the implications arising from treading on existing paradigms when considering a shift in perception. In her case, she has noted that people had a visceral response when she initiated a shift in paradigm from "cancer is bad" to "cancer is neighbor." This flies in the face of many peoples' personal experience of cancer. People perceive the shift in paradigm as threatening. Yet, as Ness points out, a different perception of cancer could lead to a broader array of useful healthcare approaches.[13] Paradigms around cancer exude strong emotional linkages as do other paradigms. When you nudge them, it's much like poking at a beehive. You may do so at your own peril. It may be better to remove the bees from the beehive or at least the stingers from the bees before planning your feast of milk and honey.

There is nothing inherently wrong with shared realities as long as you understand the nature of the game you have agreed to play with your colleagues. Many times this game can have the same entertainment value as being the nerd at the school dance. You go; you participate. You think it might not be as bad as you think, but it usually is. When you are stuck in a rut on your research, the shared realities holding those nerd folders in place become your own worst enemy.

Reality TV

A couple of interesting questions to consider are:

1. How willing are you to see your reality shift?
2. How far from the shared reality paradigm can you extend your perceptions?

Here is an example that serves as a basis for asking such questions. This year (2012), the U.S. Fox Television Network is celebrating its 25th anniversary. This is the major network no one thought could exist. Between 1950 and 1985, the US had three major networks only — CBS, NBC and ABC. The mantra was that this triad was uncrackable. No fourth network could ever get a sufficient foothold to succeed. Early attempts to do so only reinforced that this mantra had to be true. Then came the Fox Network. Whether or not you like the shows, after programming such as 24, the Simpsons, X-Files, Fringe, Ally McBeal, American Idol and Married With Children, the roadblock to a fourth network now seems silly. Not only did Fox become a viable fourth network, it became the number one ranked network for the 2007–08 television season. That breakthrough illustrates just how much and how far mantras and paradigms can change. While such changes may take longer in science, the truisms we were taught as students can fall just as hard.

In fact, the change in media has been even greater than just the shift from three to four networks. Cable, satellite, Direct TV and Internet streaming TV have changed the entire notion of what constitutes a major network. The technological changes can actually make us grateful for the solidness of science. At least there is always gravity. However, as a researcher it is useful to have one foot on solid ground and the other on what will become the new solid ground. The goal of perceiving your shifting reality is to find that up-and-coming patch of solid ground.

SHIFTING SHARED REALITIES

It is great to advocate for enlarged perceptional awareness in science, but can it be done? If so, how might you attempt to do that? For a start, we can address what is unlikely to be helpful. You probably want to avoid brute force. Using brute force when defying conventional wisdom or challenging the paradigms that buttress your colleagues' careers and livelihoods won't go well. A frontal assault on things that are core, precious and sacred is more similar to a war game than it is to clearing a path for a new scientific understanding. It should be noted that war games are one of the highest forms of drama. If you want your research career to play like "Survivor" or "Jersey Shore," then go for brute force and the war games.

An example to emphasize this point would be a lecture we attended by a distinguished scientist, speaker, visionary and author just weeks before the proposal for the new university course based on the material in *Science Sifting* was to be considered by two committees. During the lecture, the speaker bemoaned the fact that in many educational institutions the latest research that is paradigm-shifting and the non-traditional, nonlinear material is blocked from being presented because the material flies in the face of current dogma. The speaker was accurate, but only if you use two critical approaches.

1. To present the new information, you must challenge the dogma of the majority much like how the British Redcoats fought the Revolutionary War. You announce when, where and how you will show up to fight, then you line up and wait to see what happens.
2. Or you present new information by charging into the fray sticking only to the linear, "rational" tools learned via formal education and without applying any of the nonlinear tools such as those presented in *Science Sifting*.

So, how did our non-traditional, nonlinear material end up being included in a university course and textbook? We simply chose not to use those two approaches. We fought no one directly. We did not expect a paradigm-based war and we never got one. Instead, the reverse occurred.

IMMERSE YOURSELF IN NOVELTY: TRAIN YOURSELF TO SEE NEW THINGS WHEN THEY SHOW UP

A recent study provided intriguing evidence that the conditions under which you view something determine whether you ascertain that it is familiar or entirely new. A bias exists in your ability to make the distinction based on whether or not you have recently experienced novelty. Researchers at Columbia University examined the memory state prior to being exposed to different images.[14] They found that if you have primarily been relying on historically-based memory, your chance of recognizing differences or novelty was reduced. In contrast, if you have been encountering novelty on a regular basis, you are better prepared to recognize differences and anomalies when they appear. The researchers attribute this gap to the difference between simply completing a long-standing pattern vs. starting a new pattern of information. This difference is important to us since one of the goals of *Science Sifting* is to increase our capacity to recognize and incorporate new patterns of information.

THE NOVELTY EXERCISE

1. Notice in detail the organization of your office, room, apartment or home.
2. At the end of your day, take a nonessential item (not your asthma rescue inhaler) and move it to a location where you would not normally place it, though it is good if it is in plain sight.
3. In the morning, see if you notice the item in its new location.
4. Continue to move the object on a regular basis or choose a new item to shift. There should always be something new for you to notice.
5. Pay attention to your ability to pick up anomalies elsewhere. As you continue rearranging objects in your personal environment, do you find you notice differences elsewhere with greater frequency?

SUMMARY

Your perceived reality is really an approximation of what is actually there. Developmental programming and years of training in a narrowing disciplinary focus only serve to broaden the gap between your actual vs. perceived reality. The Three Blind Mice archetype standing for — Context, Historical Use, and Selective Attention — are the hurdles to overcome in achieving a broader perceptual awareness. Yet, there are several tools that can help to expand your awareness including a return to the childlike pre-programmed state. In this chapter, we suggest that you practice moving out of your expectations for sameness and start to deliberately introduce an element of novelty into your week. We also suggest you avoid using approaches that would lock you into the status quo (assuming you are not in love with the status quo). The remaining chapters in this book are geared toward helping you move beyond the status quo by revamping your mental desktop folders for improved information access, retrieval, and use.

Chapter 8

Body of Work: Managing Your Greatest Tool

They didn't even know what they "knew." — Richard Feynman[1]

There is more reason in your body than in your finest wisdom. — Freidrich Nietzsche[2]

I speak two languages, Body and English. — Mae West[3]

In the prior chapter, "Perceiving Your Reality," we asked the question, "Do you know where you are?" This was presented to stress the importance of your being able to identify yourself in the context of your perceived reality. Self-identification is rather important. Now we want to extend this question to ask "It's 10PM, where are you in relation to your body?" Of course we might have to back up a step and first ask "Are you even in your body?" Many of us spend a considerable amount of time with our awareness on some world tour rather than in our bodies where it can be most useful. Asking if you are in your body and, if so, where in your body is central to this chapter's topic: the role the body plays in creative thinking.

Fully occupying your body with your attention can be very useful, particularly if you are planning to pursue anything other than the most rote work. If you conclude that you are in your body then a follow-up question is "What do you know and where do you know it?" You may have immediately answered, "In the brain of course." But are you really that sure that your answers are restricted to your brain? We hope to persuade you that visiting only your brain for answers may be grossly underutilizing the information that your body has available for problem solving.

As we will discuss in the following sections, the body is a powerful tool for: (1) recovering useful information when you need it, (2) making novel connections between patterns of information, and (3) overcoming road-blocks in research. In this regard, we will consider, if only briefly, your entire body.

EMBODIED COGNITION

The Stanford Encyclopedia of Philosophy defines embodied cognition as follows: "cognition is embodied...when aspects of the agent's body beyond the brain plays a significant causal or physically constitutive role in cognitive processing."[4] The emphasis is on the beyond-the-brain role of the body in playing a significant role in cognitive processing. It is the science behind song lyrics like Earth, Wind & Fire's "I can feel it in my bones."

Some proposed effects of the body on cognition are its actions as:

1. Constrainer

Constraints can introduce perceptual biases

2. Distributor

Distribution allows non-brain and non-neural parts of the body to play a role in cognition

3. Regulator

Regulating functions of cognitive activity affects perceptions involving space-time.[4]

One variant on the theme of embodied cognition is something known as grounded cognition, which was examined in a review of the same name by Lawrence Barsalou.[5] Grounded cognition also leaves open a role for the beyond-brain body in cognition. From Barsalou's perspective, cognition is grounded many ways. Among these are: simulations, situated action usually among groups, and bodily states.[5] Of course this immediately begs the question: what bodily state(s) will lead you to effective problem solving? It is a question worth asking as the dividends of being able to use your body as a problem-solving asset are potentially great. We should note that while many cognitive scientists embrace the concept of a beyond-the-brain role for cognition, there are some who remain skeptical.[6]

If you have ever wondered whether we can "think" beyond what we assume to be the normal brain-centric cognitive pathways, the answer appears to be "yes." There is more to the story than meets....the brain. Indeed, our bodies are very much connected to the environment and perceptional patterns of information, and these play out in both the physical and emotional realms. Much of the research is relatively recent on this topic, and it is not our intention or desire to wade deeply into the psychological sciences. We merely provide an introduction to the topic given its potential importance for creativity in research and technology.

The concept we would like to consider is the possibility that the state of your body and its interactions with the environment can influence how you perceive and judge scientific information. Obviously, being aware of these often hidden cues can be critical for approaching research topics with desired impartiality. Additionally, once you are aware of these body connections to environment and the critical role they can play in cognitive perception, you have a new tool for affecting your perceptional status in ways that best serve your research efforts. As you will see, your body's state to optimize observing important anomalies could be as simple as what beverage you choose to drink as you review the latest data.

In a recent review, Isanski and West[7] described the basis for our use of sensory-rich metaphors and what that means in terms of mind-body connections to patterns. There are underlying links between physical states, colors and emotions that, if not fully appreciated, can influence our perceptions and scientific evaluations. The authors explain how metaphors such as "forward-thinking," "take a step back," "cold-hearted," and "grounded perception" all have a physicality that our bodies play out in real life.

One of the connections that exists is between perceived temperature and the level of trust or degree of interpersonal warmth. Zong and Leoardelli[8] showed that perceptions of cold and social isolation go together. Groups experiencing social exclusion reported perceiving that a room was colder than individuals who were socially interactive with others. Socially-excluded individuals also requested more warm food and drink compared with those who had been socially included. The reverse association also occurs. Holding a warm drink can influence how generous you feel toward a target person.[9] Similar associations have been reported in additional studies from Yale University.[10,11] You can envision the ramifications of these findings. You are grading papers or reviewing a grant proposal while holding either: a hot cup of coffee or a cold iced tea. Have you been aware that the temperature of what you are holding could affect your relative perception of the information? In reality, the effects of this single body cue may be subtle. But as cues multiply, their added effects could well influence the course of your day or week.

Another metaphorical body-environment connection is between time and space (as direction of motion). Miles *et al.*[12,13] found that mental travel through time (i.e., remembering the past or envisioning the future) is associated with distinct patterns of body movement through space. Forward movement is connected with thoughts about the future and backward movement with remembrances of the past. This association holds for both directions regardless of whether: (1) you think time-related thoughts and the body synchronizes them with forward or backward motions or (2) you initiate a rocking motion only to realize that your thoughts are now meandering across future or past events.

Probably the most debated area of extended cognitive science concerns what is known as the Extended Mind Thesis.[4,14,15] This is the question that if our body is involved in cognition, is all of this cognition exclusively within the physical boundaries of the body? Can cognition be extended beyond our physical body? The perspective of Wilson and Folia[4] is that a core of cognitive scientists probably embraces the narrow physical body-bounded view of cognition, but a group of scientists suspects that cognition extends beyond the physical body.

Finally, it is worth mentioning how embodied cognition is applied outside of science and research. As you might imagine, in business when there is money on the table, it is a matter getting it right and not just a matter of how your beverage might affect how a grant proposal is scored. Team managerial efforts in business often involve time-sensitive interactions among groups of people with significant financial implications. This may be one reason why the University of Manchester's Embodied Cognition Lab is an interdisciplinary effort between the School of Psychological Sciences and the Manchester School of Business. An examination of whether business has adopted embodied cognition as a reality and a useful tool suggests it is the *modus operandi*.[16]

CREATIVE VISUALIZATION

Creative visualization has been given many terms including visualization, guided imagery, and mental rehearsal. It is the process that allows you to imagine actions or behaviors.[17] There is evidence suggesting that by imagining yourself doing an activity or performing a specific function, your body thinks you have actually performed it. One of the first applications of creative visualization was in sports.[18] Athletes would imagine themselves successfully engaging in sports activities. One example might be shooting and making free throws in basketball.[19]

In another example, Ranganathan *et al.*[20] at the Cleveland Clinic in Ohio examined three groups of 30 individuals. In group 1 (n=8) they had the individuals visualize they were exercising the abductor muscle in their little finger, group 2 (n=8) was doing the same in their biceps muscle and the group 3 controls (n=8) visualized no exercise of either muscle group. Visualization lasted 15 minutes, 5 days a week for 12 weeks duration. Another group of six individuals performed actual physical exercise of the little finger muscle for comparison. Muscle strength was measured before, during and after the visualization training. Muscle strength improved 53% in the physically exercised group (finger) with no change in strength for either muscle in the group 3 (controls). After visualization, Group 1 increased finger muscle strength by 35% while group 2 had an increase of 13.5% in biceps muscle strength. The findings suggest that imagined exercise of muscles significantly increased their strength though not to the same extent as actual physical exercise.[20]

In a more recent study, Reiser *et al.*[21] studied the effects of imagery of the muscle contraction, which they termed imagined maximal isometric contraction vs. actual isometric muscle training. This comparison was assessed using four strength test exercises. They found that the imagined muscle contraction, training exercises could at least partially replace actual muscle training without appreciably reducing the amount of strength gained.[21] These various studies suggests that at least for some endpoints, envisioning a body activity can have a very similar result to actually performing the activity. But if repetitions of muscle activity can be envisioned instead of actually being performed, can a higher standard of body-directed (muscle) performance also be envisioned?

Visualization — One Experience

Since the authors have personally tested out most of the tools in this book, including of course all the toys, you may be wondering if we have an example of personal visualization. We do although keep in mind that it is only a personal experience. The example comes from tennis. I (RRD) had enjoyed playing tennis with a university coworker whenever schedules and weather permitted across eight consecutive Ithaca summers. My coworker was very gracious to continue to play each summer. Given that we played approximately 7–8 times each summer, my record was about... 0–60. If the number of losses was an approximation, the number of my victories was quite precise. That was until I decided I would try visualization once.

I should indicate that mine was a variant on the standard visualization. Instead of imagining myself serving and volleying at my own best or doing it repeatedly over and over until I eventually saw myself get better, I went for the easy route. I saw myself playing more like Roger Federer. I also had this frame of reference well-engrained since my younger son and I had been privileged to see Federer play at a prior U.S. Open tournament, which he won. I visualized myself playing more like Federer than my usual self. The next match after the visualization, I played my best ever and I won. So my record is now 1-something over 60. It was only an experience but a useful one. Because in the end, it could just as easily have resulted in a new scientific discovery rather than a different outcome between two middle-aged guys getting a little exercise. Next time I might try to visualize Barbara McClintock examining maize plants or Richard Feynman eating lunch. Keep in mind, I do know where McClintock's maize field was located and where Feynman ate lunch on our campus.

BODY POSTURE, MOVEMENT AND PATTERN ENGAGEMENT

Movement

We might have titled this subsection on movement, the "hokey pokey" (as in the dance). That is because it involves not only the topic of posture but also various body motions (of which the hokey pokey dance covers many). The idea of connecting both body posture and movement to our environment is not a new idea. We tend to lean forward toward things we desire (e.g., an appetizing dessert vs. a rock, except for geologists). However, recent studies have demonstrated the extent to which our bodies provide a rapid response to changing environmental stimuli. For example, Eerland *et al.*[22] found that undergraduate participants leaned to the front when engaging pleasant images and to the back with unpleasant pictures. The avoidance response was slightly delayed compared with the initial approach response.

Side-to-side motion also has potential interactions with the environment and our perceptions. For example, University of Bern researchers reported that body movement directed to the left or right was intertwined with the processing of numbers.[23] The authors concluded that body motion is involved in higher-order spatial cognition. Rotational motions can also be involved in pattern engagement. When participants were asked to read maps, it was found that they performed mini rotations of their head that were related directly to the main map reading task. So for front-to-back, side-to-side and spinning yourself around (or at least your head), the hokey pokey is part of your interaction with informational patterns and higher-level cognition. Body movement has been suggested as a route for improved education of children.[24]

Here is an exercise to try that will help you to calibrate where your body is in relation to your environment.

THE BODY-WEATHERVANE EXERCISE

(don't do this exercise immediately before a crucial work deadline)

1. Surf the web (preferably seeking local area news sites) and find the three most horrific news stories of the past week or month. Crime or disaster stories are the best for this and our news is filled with such stories on a daily basis often with graphic photos or videos. Read the first paragraph of all three stories and the entirety of at least one of the three stories. Look at any and all pictures accompanying the stories.

2. Notice any change in your body. Take this calibration. Your respiration may be quicker. You are likely crouched forward with your shoulders pulled in toward your body, your chest pulled in and stomach pulled in as if protected. Your eyes are forward and probably your peripheral vision is reduced. Your body takes the stories to heart and actually postures you for fight or flight. You are more into the stories than you might have imagined.

(Continued)

THE BODY-WEATHERVANE EXERCISE *(Continued)*

3. Now, stand up and walk around much like a baseball batter stepping out of the batter's box after being fooled on a pitch and having a really bad swing. You and the batter are hitting the reset button. If useful, find a good double metaphor to reset your brain. If it is needed, take a longer break before continuing.

4. Sit back down and upload a video from your computer or use a video site such as YouTube of your favorite band playing one of their songs or music videos.

5. Now calibrate where your body is. How is your respiration? How does you posture compare to where you were after reading the news stories? shoulders, chest, stomach? How is your vision (direct and peripheral)?

It is quite likely that you noticed a significant change in your body between the two experiences. If not, you may want to try a different set of news stories and music selections (i.e., if your music choice spurred you to sing a Beatles song at top voice or try dancing salsa for 10 minutes even while seated, your respiration may be elevated). In that event try alternative music you like that does not automatically throw you into dance or Karaoke mode).

Watch for these body signals as you go through your research day. They can help you track the factors from the environment around you that may be aiding or inhibiting those moments of creative inspiration. Next in this chapter, we will consider a related tool to enhance creativity: deliberately changing your body posture and state to recreate the open and relaxed "favorite music state" in your body and enhance your creativity.[25]

Posture

Body posture itself can also play an important role in your interactions with patterns of information. Your own posture as well as that of others can affect your perception of patterns.[26] There is evidence to suggest we do our best to take a holistic view of others including body posture. Mondloch[27] found that both children and adults took longer and made more errors in evaluating facial emotions depending upon whether body posture was congruent with the facial emotion or incongruent. If the body posture we perceive in others is holistically inconsistent, we are confused by the presenting pattern. When the posture pattern is incomplete, we will try to use our own posture-derived information to fill in missing information. Following experimental studies of participants interacting with a face and hands that could represent different body postures, Kessler and Miellet[28] suggested that the observer intuitively uses his or her own body schema to "fill in the gaps in the stimuli."[28] Keep in mind these stimuli are simply patterns of information.

As an observer your body will physically complete a pattern that you have engaged. The authors speculated that body-gestalt completion of postures may be a social and survival mechanism to detect deception. Consistent with the idea of a sideways body motion and numerical quantities, posture seems to follow a similar relationship.

Recently researchers at the Open University in the Netherlands won the 2012 Ig Nobel Prize Award for psychology in a ceremony held at Harvard University. The Ig Nobel Prize is awarded for science that first makes people laugh and then makes them think. That is a category deserving special recognition in *Science Sifting*. This year's awardees had the intriguingly-titled paper, "Leaning to the left makes the Eiffel Tower seem smaller: posture modulated estimation."[29] In this study, Anita Eerland and colleagues asked participants to make numerical estimates. But they surreptitiously caused participants to lean to the left or to the right (compared to being upright) using a Wii board. The leaning posture caused the participants to underestimate (leaning left) or overestimate (leaning right) in their answers to a series of questions involving numerical quantities. The participants were not aware of this experimental effect.[29] This could be useful to know as you contemplate the relative magnitude of your latest research findings.

Another important aspect of body posture is not just what it can do to affect current perceptions but also how it affects information retrieval. Dijkstra *et al.*[30] showed that when you attempt to retrieve memories, if you assume a body position that is similar to the one you were in when you first encountered and accessed that pattern of information, it will be easier and faster to retrieve the same information from your memory banks. The time interval they employed in the experiment was a memory from two weeks prior. The ramifications of this are that using more than one body position/posture may allow you to bring forward different sets of previously obtained information as you grapple with potential interpretations of research data. At least that is one explanation you can have on hand when someone arrives at your office to find you attempting a headstand.

BODY-CENTRIC THERAPEUTICS AND EMOTIONAL PATTERN RELEASE

Have you ever gone on a trip and realized before the end of the first day that you have over packed and will be burdened with hauling excess baggage every place you go on your travels? Nowadays you are probably paying extra fees for your extra baggage at the airport, taxi and hotel. But it is also possible you haul extra body baggage into offices and labs that slows you down, biases your perceptions, distracts your focus, and clutters your informational folders. You may not even be aware that you are doing this. Before this starts to read like some kind of advertisement, the point we are making is that informational patterns can have an embodied component and many of those patterns are connected to our emotions. If you like the idea of occasionally cleaning your home, office, and lab, you might want to consider the benefits of doing the same thing with your body when it comes to some well-ensconced, emotionally-laden patterns.

One of the major growth areas of alternative and integrative medicine surrounds the concept that by shifting physical aspects of the body, emotional patterns can be released.[31,32] This is such a common phenomenon that entire practices have emerged geared for this sole purpose. We would emphasize that patterns of information are really not so different beyond how and where we experience them in our bodies. In chapter 12 we describe how a pattern of Faust's monolog connected to Tesla's hand for whatever reason. A pattern of information that includes your anger from an earlier-life incident may be connected to your spine. If you still have doubts that informational patterns go beyond our brain and have an embodied component, spend a day in the office of a practitioner performing acupuncture,[33] chiropractic techniques,[34] cranio-sacral therapy,[35] deep-tissue massage,[36] emotional freedom technique (EFT),[37] myofacial massage,[38,39] neuro-emotional technique (NET),[40,41] or somato emotional release.[42,43] Often body work practitioners observe clients releasing emotions as they work with them. Of course shifts in embodied patterns of information unconnected to emotions can occur as well but may be far less obvious to a third-party bystander.

Personally, we have used a technique, called *source field guided tours*, that combines meditation with sensory-awareness-and-visualization to explore patterns of information in the terrain of the body. The guided meditative body tours often produce emotionally-laden awareness and/or release as the body is navigated. The process is much like visiting a massive multi-floor department store. Each level of the body is stocked with different landscapes and some of them connect to emotional patterns (including sheer joy should you find the right floor).

Being able to identify, connect with and shift out the unhelpful patterns of emotionally-based information held in your body is much like an annual cleaning of your lab. You discard outdated chemicals, reorganize the workspace so it is more efficient, welcome in new shipments of reagents and get ready to add new staff and a new vision to your research effort. For this reason, we highly recommend a periodic body tune-up clearing the way for your innovative research initiatives.

SUMMARY

When you interact with patterns of information, your body is a full participant. It responds to your interaction with the external environment. Even if you are not aware, your body adjusts as you engage in social interactions to permit you to better comprehend and to feel what someone is telling you. Your body can be a key to accessing memories as its position can impede or facilitate memory retrieval. Your body posture can signal the openness of your perceptional awareness as well as your receptivity to new ideas. Directional motion of your body has actual meaning reflecting time, space and numerical relationships. The relationship of thought and body position are bi-directionally linked. Think about something and watch your body assume a position. Assume that position and you are likely to think about the linked category of thoughts. Armed with an understanding of embodied cognition and these relationships, you can begin to use your whole body to break through roadblocks. In fact, visualizing your whole body moving you past a roadblock is not a bad place to start.

Chapter 9

Creative Spaces

Imagination is more important than knowledge. Knowledge is limited. Imagination encircles the world. — Albert Einstein[1]

Be brave enough to live life creatively. The creative is the place where no one else has ever been. You have to leave the city of your comfort and go into the wilderness of your intuition. What you'll discover will be wonderful. What you'll discover is yourself. — Alan Alda[2]

Your creative space is a mental playground where you go to create solutions...but...most of us are socialized out of childhood's joyful creative spaces. — Gerald Nadler and William Chandon[3]

One of the important concepts in this book is the idea that, on a daily basis, scientific researchers can purposely design operating spaces that allow for improved pattern interactions, better pattern recognition and more creative thinking. We term these: creative spaces. Creative space is not about enlarging your research office space (after all, mine is a state-mandated maximum size). It is about gaining personal, and in some cases, communal permission for you to become your true creative self. There are excellent exercises to bypass the roadblocks to creativity and innovation. However, it is very helpful to give yourself permission to do what your left brain may well resist doing. By opening up a creative space, you intentionally move yourself into nonlinearity.

FENCES CAN MAKE FOR STALE RESEARCH

When we first set out to discuss the topic of creative spaces, an old Cole Porter song kept running through my head. It has the lyrics "Oh give me land, lots of land, under starry skies above. Don't fence me in…." While no longer exactly a top 40s hit, it epitomizes the overall theme of this chapter. You need to remain free of fences, unbounded by personal filters, biases, others' expectations, your prior experiences, and possibly even your own comfort zone in order to operate fully in creative spaces. Operating in creative spaces can have consequences since not everyone around will be doing the same. But the rewards of discovery born out of creativity can be most satisfying and useful. This chapter discusses strategies to create and hold open creative spaces to encourage innovation.

Moving beyond the fence

Creative spaces are a little like a void or a holographic template. Once the space is created, it demands to be filled with something. One tendency is to define your creative space and then just leave it there. This can make for a succession of research days much like the Bill Murray experience in the movie, "Groundhog Day." You are likely to see repetitious sameness hoping for that one useful deviation. Deviations can show up but only after a lot of energy-intensive repetition. Additionally, aging spaces inevitably acquire the clutter of status quo. This is likely to dominate your awareness, focus, and energy. As a result, a more useful strategy is for these spaces to be created anew each day. It is not a matter of telling yourself "perhaps I will be remarkably innovative today." It is a matter of being open to new possibilities each day. How open are you to something different showing up? Have you given yourself permission for your paradigms to be shattered?

ROADBLOCKS: BOTH PERSONAL AND COMMUNAL

In *Innovation Generation,*[4] Roberta Ness describes two types of impediments to innovation: frames and biases. These impediments color the way a researcher takes in, organizes and reaches conclusions from the data. In our terminology, Ness' frames and biases would be termed folders and memes, respectively. Our folders affect the spectrum of data we allow ourselves to consider and whether or not we screen out data because we have no previous reference for it. Memes are cultural or behavioral restrictions that we share with others (non-genetically) and that color how we see the world. We will discuss these in more detail in another chapter. Folders and memes can be highly individualized. But they are also impacted by the environment in which we are immersed. Because we are so interconnected with our environment including those we interact with, it is important to consider how roadblocks may be connected to your surroundings.

You may think that logic would dictate that we should begin by considering the individual scientist and then work outward to broader surrounding spheres of influence. Of course, since we are moving deeper into the more nonlinear parts of this book, we will take a different approach. After all, part of this chapter is about swapping the precisely logical path for more possibilities, about breaking paradigms, and about turning things on their head.

It is easy to think that by addressing our own personal roadblocks and creative spaces, the outside world will simply take care of its own space. But our experience is that personal and communal space go hand-in-hand. They can tend to mirror or feed off of each other. Roadblocks held partially in place by communal traditions can be easily swept aside if your efforts to open creative spaces include a healthy infusion of creativity for the entire community. It becomes much easier to enlarge and maintain your own creative space when those in your immediate surroundings are playing the same game.

This is depicted in the following diagram (Figure 9.1).

For this reason, we will dive into our discussion of creative spaces by talking about the research community. That is the scientist and his or her relationship to the neighborhood or community of those with whom we interact in professional life: other scientists, coworkers, students, administrators, lab members, clinicians, business partners, governmental agencies and foundation staff, etc. As a current scientist or a researcher in training, you are unlikely to operate in total isolation. Unless you are the research equivalent of a one man or woman band, the dynamic of interactions with others can, to some extent, affect your propensity and success at holding open creative spaces.

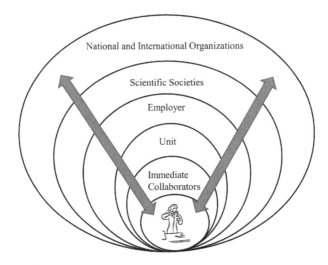

Figure 9.1   Interactions Among Personal and Communal Creative Spaces

TIDYING UP THE NEIGHBORHOOD

The neighborhood is a designation we give for your physical surroundings and interpersonal interactions that can influence your opportunities for creative thought. How is your neighborhood today? What kind of neighborhood do you hope to be a part of in your own pursuit of scientific research? The ways in which we define, form, and use neighborhoods in our day-to-day operations can greatly affect our capacity for creativity. While the work will be your own, creative or not, your neighborhoods can either facilitate creativity or make your efforts to move beyond conventional wisdom a Herculean task. They either impose a template for uninspiring and mundane outcomes, or they pave the way for creative ideas and innovative pursuits.

You may not realize it but you are already part of one or more neighborhoods. These multiple neighborhoods overlay each other to form your scientific world. You already received the invitation much like in the song lyrics: "It's a beautiful day in the neighborhood... won't you be my, won't you be my... neighbor" (adapted from the Mr. Roger's television show). You probably joined the neighborhoods quite passively and may have done little to actively manage them to the creative advantage of both you and others.

Neighborhoods encompass your scientific colleagues, students, trainees, bosses, scholarly societies, funding organizations, speaking and publishing venues. They establish what is expected, what is considered highly useful and productive, what is considered unimportant or a low priority, and what is deemed innovative, "cutting edge" or, in contrast, unworthy of pursuit. The neighborhood is like a bubble universe unto itself. You have entered it, and it can be very difficult to see outside the bubble. Obviously, if it is hard to see beyond the bubble, then venturing outside would be high risk as you would move into unknown and potentially dangerous spaces.

So what might be limiting about a neighborhood? How could it restrict your creativity? Well, history tells us that a majority of innovations were first seen by the "neighbors" with opinions ranging from disinterest and disillusionment to frustration, disdain and even outright anger. Terms such as: not useful, out of focus, a distraction, impractical, beyond the pale, unsupported by current thought, high risk, has no future, a waste of time have dotted the landscape of innovative breakthroughs. The innovator has come into a neighborhood where state-of-the-art, cutting-edge contributions really amount to polishing the apples already sitting on the apple cart at the market. In actuality, it is not very thrilling despite the community hoopla and accolades that may come as a result of apple polishing. Instead, the innovator invariably disrupts existing paradigms or even rocks the world of some colleagues. He/she has effectively upset the apple cart (and certainly some of the neighbors along with it).

I (RRD) will digress to describe a personal professional experience with the hope it will highlight what upsetting an apple cart might look like. One of my personal benchmarks for butting up against an existing paradigm occurred at a national invited lecture I gave in Washington, DC. in the Spring of 2005. It was an experience that I will never forget. It was one of the first times that I directly challenged the decades-long dogma surrounding what constituted adverse immune outcomes from harmful environmental exposures (e.g., to chemicals and drugs). From the 1970s to the current decade, the prevailing paradigm had been that environmental exposure producing immune suppression and increased risk of infections (and possibly cancer) were the greatest immune-related health threats. The HIV/AIDS epidemic did little to dispel this notion. As background,

I should say that my own research and that of others beginning in the mid-1990s led me to the conclusion that harmful, early-life, environmental exposures were part of the basis for the recent epidemic in allergic and autoimmune diseases. In effect, that immune suppression was not the be-all concern and probably not the highest priority present day concern (based on CDC health statistics).

So in that spring 2005 lecture, I detailed a model in which problematic outcomes for the immune system (known as "immunotoxicity") were not limited to immune suppression (then the prevailing dogma). Instead, I explained how toxic chemical exposures could also increase the risk of allergic and autoimmune responses (improperly enhanced immune responses that produce allergic and autoimmune diseases). Remarkably, before I could even leave the podium, one of the lecture attendees ran down the aisle and started screaming at me. I should add that this was not the equivalent of audience screaming one might have heard at an Elvis, Beatles, Bon Jovi, Bruce Springsteen, Justin Beiber, Psy, Beyonce or Lady Gaga concert. Instead, she was telling me that "I had no right to say that immunotoxicity included allergy and autoimmunity. It was ONLY immune suppression."

Despite being shocked at being yelled at (my previous experiences probably included loud snoring but no yelling), I found this slightly ironic. After all, my academic title at Cornell University (then and now) is "Professor of Immunotoxicology." In fact in 2005, I may well have been the only person in the U.S. if not the world with that title. Yet, I was told I had no right to define the area of immunotoxicology as such. The implications were clear enough. If immunotoxicity from environmental exposures was defined to include elevated risk of improper allergic and autoimmune responses, then childhood asthma, type1 diabetes, lupus, rheumatoid arthritis, celiac disease, food allergies, autoimmune thyroiditis, inflammatory bowel disease, multiple sclerosis, and countless other epidemic-level chronic diseases were likely outcomes of immunotoxicity. No longer would immunotoxicity carry the non-descript generic label of immune suppression. This new model was in effect, a game-changer.

At the time, I wondered if I needed to get measured for a bullet-proof vest before my next lecture, but then a remarkable thing happened. Not all at once. Rather, in a slow yet steady progression, the neighborhood began to change. It was as if a dam burst after the yelled-at lecture. First it was a slow trickle, but then literally a gushing forth of more open thinking that was threatening the status quo dam. In hindsight the yelling scene was a symbolic communal turning point. Now only seven years later (relatively quick in paradigm-standard time) and after a few less contentious lectures and publications, the World Health Organization[5] has prepared a definitive monograph that guides risk assessors and public health officials in the landscape of chemical-induced immunotoxicity. In addition to immune suppression, it includes as formal categories: risk of allergy, autoimmunity and improper inflammation. In 2005, I head-butted a paradigm and there was push-back. But in 2012 there is a new paradigm.

So, are scientific innovators destined to always swim upstream and to be as popular as Frankenstein's monster showing up at the village's marshmallow roast? Not necessarily. While innovation comes from the individual scientist, the neighborhood can either: 1) represent a massive problem or 2) be a highly-supportive vehicle for innovation. To get at neighborhood resistance vs. openness goes beyond teaching individual students how to be more innovative. It is the web through which recognition and acceptance of the keys to innovation extends to broader and broader communities of scientists.

TEST

Here is a test. There is no right or wrong answer. In fact some level of confusion may be the most useful outcome.

Background: Most scientific researchers learn early-on that there is an important stratification among peer-reviewed research journals. The journal in which you publish your research and scholarly ideas matters. It can determine how your research is viewed by colleagues and can affect how it is perceived in terms of significance. This is a little like real estate, location matters. Some journals are considered exceptionally prestigious and are badges of honor for your research publications. You are likely to try to place your best work in those journals although competition is fierce and acceptance rates for papers may be low. Others are quality journals but are more likely to attract your

(Continued)

TEST (*Continued*)

average research papers (e.g., quality negative-results studies). Those are papers that may move things along but perhaps in baby steps. Still other journals are likely to be beneath your radar. These might represent either new journals starting to make a name or journals that tend to publish lower-quality research papers. Journals can even be quantified based on citations of the papers they publish. This is the numerical "impact factor." Research faculty evaluations are often based not only on the number of papers published per year but also on the impact factors of the journals where those papers appear. When it comes to tenure evaluation, a few publications in highly prestigious journals may trump many more publications in low impact journals. Again, this is what academic researchers know as the standard operating procedure.

Question: Based on the history of innovation, where are the most innovative research concepts likely to be found: a) in the most prestigious journals, b) in more easily accessible journals that tend to publish lower quality papers?

Answer: Quite clearly the answer is b. The most novel ideas that shatter paradigms are unlikely to make it through the review process of the most competitive, top flight journals. For example George Kreb's Nobel Prize-winning research on the citric acid cycle was rejected by the journal, *Nature*, only to eventually be published in the less prestigious journal *Experientia*.[6] If these ideas are published at all, their first appearances are likely to be in a lower echelon journal. Now as a researcher who supports the neighborhood of scientific innovation, where will you focus your reading time? It almost certainly has to be on the most prestigious journals. You have to know the content of that literature to even to talk to your colleagues over tea or coffee. But you can be a helpful contributor to a neighborhood that supports innovation by recognizing that those journals are rarely the initial starting point for the most innovative, paradigm-shattering scientific ideas.

There are several questions to ask about creativity-supporting neighborhoods. How would you go about encouraging creativity-supporting neighborhoods and communities? Is this something we have to create anew or are there already models for such community support? How do we form and expand creative spaces in one scientist's office and then in the whole neighborhood? What might a community holding creative space look like?

We can use historical examples of impressively high levels of neighborhood creativity as one source of potential answers. While there are many ancient lengthy examples, one more recent example is the era of Scottish Enlightenment and Exceptionalism that had its origins in the 18th century. Of course, the irony is that for many of those who participated in the inexplicably innovative period of Scottish history, it may have been perceived as the status quo.

THE SCOTTISH EXAMPLE

Scotland itself is interesting as historians have characterized it as being relatively poor and having a comparatively barren terrain. Yet, the sheer inventiveness and creativity among its modest-sized population was remarkable. It is almost inexplicable that so few people going about their daily activities so affected the world. Some historians see the inventions literally flowing from Scotland particularly during the 18th and 19th centuries as giving rise to what we call "the modern world." Of course others might see this as an overreach in characterization. Regardless, the period bounded by the 18th-early 20th centuries was a historical "hot spot" of Scottish innovation both in Scotland itself and via her emigrants. The following examples may provide a glimpse of the keys to this creative burst.

Money, Money

Although there were substantial inventions before the 18th century (e.g., Napier's discovery of logarithms), we can begin in the early 18th century with John Law of Lawriston. Law was part of a distinguished family of Edinburgh goldsmiths. His wide range of interests and belief in his own ideas led him to revolutionize banking through the creation of currency as it is known today: paper money. At one point in his career, Law, then exiled from London for his swashbuckling ways, became the controller-general of France and literally controlled all of that country's money. By tightly linking banking and commerce, Law's reach extended around the globe to influence most trade between Europe, Asia and South America. Law so controlled world finances that when an investment bubble he had promoted spectacularly collapsed, it literally jeopardized the regent of France and that country's stability. John Law simply operated outside of what were then the perceived boundaries of economics as well as (from his London exploits) the law. In fact, remarkably he was so respected that even after causing a world-wide financial disaster, he was a welcomed dignitary in several European countries.

Enlightenment and Innovation

Other examples of Scottish creative exceptionalism from the 18th–20th centuries can be found. David Hume[7] is often credited with helping to usher in the era via his work in empirical philosophy and his advocacy for a world view among his countrymen. Ironically, it was Hume who also led the charge to purge Scotland of colloquial linguistics that might impede the impact of Scots on the world. As detailed by Brown and Dietert,[8] there was a general loss of Scottish identity that coincided with several 18th century events: 1) the Union with England (1707), 2) the Jacobite defeat at Culloden (1746) and 3) Hume's treatise on the purging of Scotticisms.[7] One feature of collective creative space could be summarized as follows; it includes a collective loss of cultural/personal identity. This is the equivalent to the loss of community memes. However, the apparent "loss" is a bit of an illusion since it is really an opening for new vistas and opportunities. At the community level we would describe this as removing the clutter of the status quo.

What happened in Scotland and among Scots who emigrated to other countries was a blizzard of discoveries and inventions particularly during the 18–19th centuries. These included: the theory of electromagnetism (James Clerk Maxwell), modern economics (Adam Smith), print stereotyping (Edinburgh goldsmith, William Ged), universal standard time (Sir Sanford Fleming), development of the historical novel (Sir Walter Scott), the science of geology (James Hutton), the revolution of neoclassical architecture (Robert Adam), the hypodermic syringe (Alexander Wood), the threshing machine (James & Andrew Meikle), the reaping machine (Rev. Patrick Bell), the refrigerator (William Cullen), the steam hammer (James Nasmyth), the flush toilet (Alexander Cummings), the modern lawnmower (i.e., sheepless) (Alexander Shanks), macadamized road (i.e., modern day pavement) (John Loudon McAdams), the pedal bicycle (multiple inventors), the electric clock (Alexander Bain), postage stamps and postmarks (James Chalmers), the first encyclopedia of pharmacy (William Cullen), color photography (first use, James Clerk Maxwell), the Kelvin unit of temperature (Lord Kelvin), the noble gases (William Ramsay), the process leading to oil refining (James Young), the steam engine (James Watt) and the telephone (Alexander Graham Bell). The 20th century brought the discovery of penicillin as an antibiotic (Alexander Fleming) and the invention of the television (James Logie Baird).

Not surprisingly, one of the theories put forward for the era of Scottish Exceptionalism pertains to the high level of education and significant percentage of literacy in Scotland at the time. Certainly education and experience are important. Malcolm Gladwell[9] in his book *Outliers* identified extensive experience as a useful factor but not the only basis for innovation. The exceptions to this rule can be staggering. Take for example the invention of panoramas by Robert Barker.

Why Creative Spaces are so Important

As described in a recent article by Brown and Nasilowski,[10] Barker was an uneducated itinerant Irishman who, after a business failure in Dublin, had come to Scotland with his wife and children seeking work as an art teacher and portrait painter. In 1786, he was strolling with his daughter on Calton Hill just outside Edinburgh and was gazing at the view when his daughter asked him why he was so deep in thought. He replied he was thinking about how to produce a "total view."[10] Later, on the same hill with his 12 year old son, they proceeded to frame out what became a London 360 degree exhibition of Edinburgh's city view with castle. Coining the term "panorama" from the Greek meaning "all seeing view," Barker became wealthy through mounting exhibitions that literally placed the observer into the midst of landscapes and scenes.[10] The portrait painter who had struggled to make a living by focusing on and accurately representing details of individuals had zoomed out to incorporate large scenes with astonishing results.

You might think that Barker had simply discovered a goldmine, filed a prospecting claim and proceeded to cash in his ore samples, but the trek was not so smooth. His first two showings preceding the successful London exhibition were relative failures. Additionally, the most successful artists of the time including Sir Joshua Reynolds told him to abandon the project. But it was then that Barker exhibited a feature that Roberta Ness[4] has stressed, perseverance. Barker was certain he had a good idea and that it could become a reality. In fact that reality, panoramas, transported Londoners to Edinburgh without the need for an arduous carriage ride.

To turn his panoramas into a reality, Barker (still little more than an itinerant Irishman 'on the make') first had to forge a near revolution in artistic perspective by designing a form of 'multi-perspective' in 360 degrees, as distinct from the more traditional 'central perspective' of 46 degrees that governed the production of framed paintings created for single viewers. Then, in London in 1787, Barker turned his hand to architecture, by successfully laying down in his patent application the principals that would govern the construction of panorama rotundas (with variations) for the next hundred plus years. And finally, the success of Barker's enterprise depended completely upon selling tickets to the public. He was personally instrumental in designing the panorama content (think of today's movie scripts) in order to sell his tickets. His subjects were mainly surrogates for travel (for example, Barker mounted a panorama of Istanbul in London in 1802) and also military-themed panoramas to illustrate the glories of the British Empire.[10]

Barker's patent, awarded in 1787, gave him sole rights to present what could be translated as "nature at a glance." In fact "nature at a glance" is an appropriate theme for *Science Sifting*. If you can: 1) access the broadest possible landscape of information (bypassing filters and memes) and then 2) recognize the meaningful patterns of information via a glance, you are well on the way to discovery. The uneducated Barker had created a type of new reality for the observers of his panoramas.

To revert briefly to Scottish Exceptionalism, Barker's invention of the 'all seeing view' in Edinburgh in the decade of the 1780s provides just about the perfect metaphor for the Scottish Enlightenment. Barker was only one of many in that city then who 'seeing all' – both 'the big picture' and the details – forged world-changing innovations with these, their visions. Have we now, by and large, forgotten something that was generally understood in Edinburgh during that brief golden interlude? Perhaps a visualized trek up Calton Hill is in order.

One interesting note about the Barker saga is that an author inspired to detail Barker's historic accomplishment is himself, a designer who thinks outside the box and creates new realities via 3-D spaces. Perhaps it is not surprising that author and set designer, Kevin Brown, is well known for designing "faux worlds." After his work on the set of the TV show X Files, Brown went on to design mock Afghan villages used for military training. Barker's legacy of "transporting" Londoners to Edinburgh is alive and well with Brown's new holographic realities.

What are the lessons of Scottish Exceptionalism and Robert Barker? They do not describe absolutes about the cultivation of innovation. But they do provide suggestions as to the characteristics of neighborhoods that are supportive of research innovations. Such neighborhoods are likely to have three features: 1) a value of broad-based education/information, 2) a willingness to accept or at least tolerate persistent unconventional thinking, 3) a willingness of participants to lose part of one's historic self-image or ego in exchange for the discovery of new paradigms.

Now that we have discussed the neighborhood "spaces" that should surround research scientists, it is time to focus on the creative spaces of individual researchers.

DEVELOPING YOUR PERSONAL CREATIVE SPACE

Playing by the Rules

One of the remarkable and most easily overlooked challenges to greater creativity in research is how we decide to play our own research game. We hope to convince you that research is partly a game and as with most games, there are rules. Do you know the rules? Do you know who sets up the rules?

Right now you may be thinking "how dare we call research a game?" After all, research is serious business. It is certainly not a game. There are centuries of history behind scientific research. It is life, death, dreams and accomplishments. It is a never-ending quest for knowledge and scholarship, an opportunity to better society, a chance to ease the burdens of people worldwide. Of course, you would be correct. But in terms of facilitating creativity, is that really how you want to start off each day? Do you really want to begin each day by placing the burden of societal survival on your shoulders before you have even had a first cup of coffee or tea?

This is where a duality becomes important. The implications of scientific research and innovative breakthroughs can be serious. It can change how we operate (e.g., whether cigarette smoking is a widely-accepted and admired activity or a readily-avoidable major health risk). It is one of the reasons we pursue our work. But when we decide to operate in the realm of "scientific research: the serious business," we are governed by rules that are not of our own making. Those rules have embedded in them a great deal of conventional wisdom. Such rules also tend to ignore what Malcolm Gladwell calls the "outliers."[9] That is the 5% tails on bell curves where the unexpected and non-conforming exists. There is a serious field effect surrounding the serious business of research, and it is something to overcome when you are seeking enhanced creativity.

In contrast, if each new research day is your new "game," then you set the rules. This is not about control. It is about command of your space.

The Physical Work Space

It may not be shocking to learn that since delving into this area, I have modified the decorations in my work office. Many of the desk items and wall hangings touting my credibility as an expert in science or reflecting accomplishments or service are no longer visible. It is useful to ask what those decorations actually accomplished over the decades? Placing them all out in view was certainly conventional. They also defined my expertise quite narrowly. But was that useful? Instead, these have been replaced with various forms of art. The art is purposefully designed to keep me from getting too serious about the work of the day. Among these is a painting by my wife that reminds me of a pleasant trip we took to Hawaii and a photograph of "Bud the Bullfrog" by local, professional photographer, Carol LaBorie. Bud is the defender of the lily pad, a humorous reminder of what nature can offer and the epitome of the "can-do" spirit. Bud helps to point toward my future rather than simply reminding me of my past.

Here is an exercise to creatively renovate the space in your office.

THE OFFICE SPACE EXERCISE

1. Look around your desk space or office and consider what decorations are really not that helpful to you in terms of your openness and creativity. Which ones have been there forever and may not be serving a useful purpose now? Remove one or more items from view (place them in a drawer, cabinet or take them home). Above all, clear some open space.
2. Take in a handicraft, other objects, or a wall hanging and leave it in place for 2–3 days.   •
3. Calibrate how your altered work space (with the new art) is now different from before.
4. Continue to test art objects for beneficial contributions to your environment until you find the ones you really want to have decorating your research office.
5. Once you finish with the physical office area, you may want to repeat this exercise with different screen savers on your work computer(s). Calibrate the effects of this redecoration of space.

DISCHARGE YOUR SCIENTIFIC BIAS

There are many tools and avenues that we can use as researchers to remove the drama and emotional charge from our science.

The Face of Science

In a section of his book *Blink,*[11] Malcolm Gladwell describes the research findings of Paul Ekman and Wallace Friesen. These researchers created a facial recognition system known as Facial Action Coding System (FACS). As an aside, this contributes to the ever growing examples of confounding acronyms in science and academia. In fact, one of our lecture topics in the creativity for researchers course at Cornell is a section we call "Acronym Acrimony" that demonstrates the narrowing effects of parochial acronym use. Examples of other uses of the same acronym, FACS, are: Fluorescence-Activated Cell Sorting, Families and Children Studies, Formal Aspect of Computing Science and Fellow of the

American College of Surgeons, Flexible Assembly Computing System. If we consider the acronym MHC we can get: Major Histocompatibility Complex, Myosin Heavy Chain and Mount Holyoke College. For fun you can look up the more than 100 uses of the acronym, COC. In some ways it is both telling and disturbing that such acronyms tend to be used without widespread confusion. This suggests that scientists who are focused on the Facial Action Coding System must not do much flow cytometry or be involved with the Families and Children longitudinal study in Great Britain. One could argue this is a prime indication of our tendency to specialize in research and to stay confined to our small backyards. I am all for acronym busting and once did give an invited lecture at Mount Holyoke College on the topic of the major histocompatibility complex.

For the purposes of this book, FACS will only stand for Facial Action Coding System. In making countless observations and then establishing rules for face-emotional connections, Ekman and Friesen demonstrated that facial expression alone is enough to create significant changes in the automatic nervous system.[12] Their conclusion was that the information on a face is not just a signal for what is going on inside our mind; it IS what is going on inside our mind. In reporting on recent research, Gladwell[11] went on to make two critical points: 1) the face is an equal player in emotion and 2) use of the facial muscles involved in smiling (even when they are NOT smiling) can cause a person to perceive what they are viewing at the time as humorous. The latter is not surprising given our discussion of "body memory" in the prior chapter on embodied cognition.

SUMMARY

Your personal and communal spaces deserve attention in the effort to support enhanced creativity and innovation. In fact they are great places to begin your creativity sojourn. If you do little else, we suggest that paying attention to your surroundings can pay significant dividends. Redecorating your home, office and lab space to: 1) point to the future and 2) avoid giving yourself narrow labels can help. Calibration can tell you which specific changes are most beneficial. Changing your personal space is easier when your neighborhood of interactions is also "open for creative business." Be an active rather than passive player in support of the creative neighborhood. You will find that creative spaces can be infectious in their impact. A small effort can grow to result in major effects. If Robert Barker can scale a hill and find his creative future, your own panorama of innovation is sitting before you ready to be viewed.

Chapter 10

Vantage Points and Gaining Multiple Perspectives

Some see the glass as half-empty. Others see the glass as half-full.
I see the glass as too big. — George Carlin[1]

A fundamental concept in art and photography is the importance of vantage points. Your position relative to your subject significantly impacts what you can see and how you interpret the connections you perceive within the information of which you are aware. In a recent article, professional photographer Danny Eitrem observed that you "don't have to be interested in photography for very long before you start to notice that almost all photos are done from the same vantage point. To improve your photography and to produce better, more creative work, you have to change that."[2] Even more relevant, Angela Belt points out that vantage points determine the actual contents within the frame as well as how those contents interact.[3] Changing your vantage point literally changes your perception of a pattern of information and how it is interconnected. Consider the merits of buying a car after only standing in front of it and looking at it head-on. Would you do that? Yet, we do this all the time when it comes to the way we examine our problems and issues. We often look at them only one way and then continue to look at them only that way.

In *Foundations of Art and Design*, Lois Fichner-Rathus[4] describes how artists often depict objects using multiple vantage points simultaneously. She stresses that, while this is an impossibility in nature, multiple vantage points provide an important advantage. They give a more complete, holistic picture. This approach in art and design is termed "multiple perspectives." Given the significance of gaining multiple perspectives, most of the tools we discuss in *Science Sifting* are designed to help you change your vantage point and access more perspectives for a broader perceptual framework.

Multiple vantage points are useful when approaching a roadblock. The steps to unblock your path may seem obvious, but you have to wonder if the dynamics surrounding the dilemma are nothing more than a default perspective with a built-in horizon and laser-like magnification on "the problem." It is also likely to be only two-dimensional, since the immediate roadblock is often isolated from the broader scope of the research. We often have to take chunks of research to focus on in order to make tasks manageable. The reality is that remaining focused on the manageable chunks often brings us to a roadblock. Whereas, allowing for multiple vantage points and seeking broader perspectives often provide enough wiggle room for anomalies to show up.

Let's illustrate the point using professional US football. As normally viewed and played, you can usually only see a portion of the field at a time and neither the coaches nor the referees can view plays in their entirety. Instead, an assistance coach located in the press box way up high provides insights as he views plays more holistically through a set of binoculars. This is one way of gaining broader perspectives. Let's face it. Given the option of being limited to a playbook while standing on the sidelines trying to see past 6'5" linemen or having a spotter in the press box, no self-respecting football coach would choose to make play decisions based only on what he alone can see. The scope of his vision is too limited; he cannot access enough vantage points.

Take the case of attending a baseball game. We have included an image that illustrates a baseball stadium (Figure 10.1). The long arrows point down to smiley faces that designate seven different seats in different locations of the stadium. As a ticket holder to the game, you can pick one of these seats depending on whether you want to be at field level, in elevated positions behind home plate, somewhere down the sidelines or in the outfield stands. The gloves and diamonds on the field indicate seven different plays — near home plate, center field fence, down the sidelines, in the infield and in right field. There is also a possible home run at the left field foul pole. Unfortunately, if you are sitting in any one of those seven seats, you will only be able to have a reasonable view of 3 or 4 plays on the field. Your ability to see the others will be poor. By only being able to choose one seat, you are automatically locked into a narrower vantage point.

Now, what if you had the ability to occupy all seven seats at once? With all of those vantage points from which to view the field, you would have a much better perspective regardless of where the play occurred on the

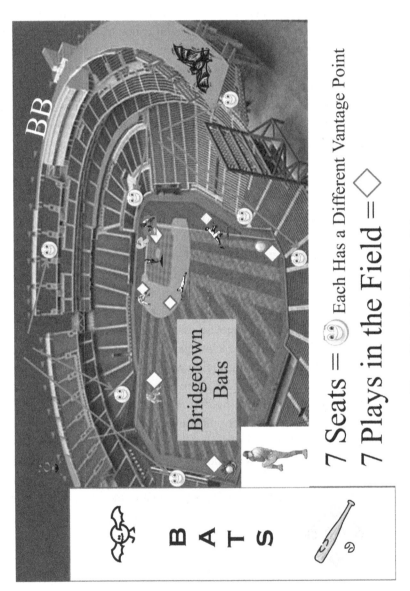

Figure 10.1   Multiple Vantage Points

field. Your view of each aspect of the field would provide equal accuracy and clarity. This is what the exercises in *Science Sifting* aim to do for you when it comes to your research and any roadblocks.

Like occupying multiple seats in a baseball stadium, multiple vantage points from which to view the stationary asymmetrical object gives you a more complete, holistic view. Now, substitute your research project for the 3-D object. How many perspectives can you glean while considering it?

VANTAGE POINT EXERCISE — THE WALK AROUND

This exercise is best done in a group but is worth the effort individually as well.

1. Take a 3-D, asymmetrical object. Place it on a table or stand near the center of room.

2. Pick one side from which to begin and mark the label "Side one" on a sheet of paper.

3. From this side, write down a description of what you see. Give yourself 60 seconds.

4. Walk right to a second point, label your paper "Side two" and again write down a description of what you are seeing from this angle. After 60 seconds, stop, and again move to your right.

5. Continue moving to your right and writing down a description of the object from that perspective till you come back to your starting point.

6. If you are working as part of a group, have everyone list their descriptions for each point. Does it sound as if you are all describing the exact same thing?

7. If you are alone, put your different sides and descriptions into side-by-side columns and look at what you've listed. How does the description of the object change based on different vantage points? Do you feel like you've actually seen the exact same object each time? Do you notice any broadening of your perceptions based on your point of view?

Now that you have completed the Walk Around Exercise, there are a couple of other methods for gaining multiple perspectives that could be useful. If your perspective was enhanced by positioning yourself around an asymmetrical object in order to gain different views, it may also be enhanced by adjusting your view based on zooming into or out from an object or pattern of information.

THE PANORAMIC VIEW EXERCISE

Remember Robert Barker's discovery of photographic panoramas of Edinburgh, Scotland mentioned in the chapter on Creative Spaces? Or imagine being an after-hours guest to the Salvador Dali Museum in St. Petersburg, FL (we actually had that experience). These are examples of experiences that can slow things down, put you into the center of the work and give you an immersive 360 degree view of what one would normally only see piecemeal. This next exercise gives you the opportunity to similarly immerse yourself in your latest research project's aims. See how your perspective of them changes as you go through the steps.

1. Take 3–6 of your most recent specific research aims and write each one down on a separate sheet of paper.
2. Lay the papers vertically on your desk in sequence. Critically appraise them and log in a notebook what you observe.
3. Now, lay the papers out horizontally on your desk in sequence from left to right and read them in that order. Do you notice anything different from this perspective as you appraise them? Log any observations in your notebook.
4. Flip the pages so they are laid out right to left on your desk in sequence and read them that way. Again, does anything appear different as you appraise them now? Log your observations. Has anything appeared to you that causes you to want to make changes?
5. Now, tape the pages to your walls so they surround you in a clockwise sequence. Stand in the middle of the room and slowly turn to view the pages as if you were viewing a photographic panorama. Does the 3-D aspect of being surrounded by your research aims alter your appraisals of them?
6. Rearrange the pages on your wall so they are running counterclockwise and repeat step 5.
7. Notice over the steps of your experience if your perceptions and appraisals of your research aims change. Notice if you felt like making changes after having seen them from multiple vantage points. How do your impressions of your research aims compare to your pre-exercise impressions? How did your impressions change between the flat, desk-top, 2-D reviews and the panoramic 3-D reviews?

Keep the pages with your specific aims!

Hopefully, you noticed things as you shifted your aims around and manipulated them in space. It could be that by forcing yourself to read the pages backward or counterclockwise, you saw anomalies that didn't come into your awareness before because your eyes were following the expected reading pattern.

The final exercise is probably the easiest to perform. It was inspired by the common observation that the majority of astronauts who go into space and see Earth from orbit have their perceptions of our planet permanently altered. Seeing Earth from such a distance is an entirely different experience than being on the ground in the midst of our world. This exercise gives you the opportunity to view your research from a similar perspective.

THE LONG VIEW EXERCISE

1. Sit in a chair facing away from your desk.
2. Take the sheets of paper on which you wrote your research aims.
3. Crumple the pages into balls.
4. Roll or toss them across the room.
5. Notice any impressions you have about your research aims now that you are viewing them from "space." Do they seem different to you? What has changed?

Even if you do not notice major impressions immediately about your research, pay attention over the coming days and weeks. Do not be surprised if your impressions begin to change. By purposefully shifting your vantage points, you have actually accessed more information than you may be aware of.

SHORTCUTS

After reading this chapter and trying these exercises you are probably thinking that you don't have the time or inclination to walk around every email message or crumple every document you produce just to obtain these additional perspectives. What you will find happening is that you will develop your own shortcut versions of these tools.

After some time of doing this, I (RRD) don't have to crumple my papers into a ball in order to get a long view profile. Now, it can happen in the blink of my mind's eye. However, the sensation of positioning my awareness in order to gain that long view perspective came from the initial physical experiences. An analogy would be that you do not have to repeat every tennis lesson on serving to be able to go out and serve. You have internalized the experience of the lessons on serving. Getting more vantage points in a deliberate, physical way works much the same way. Once you have the experience a couple of times, your body internalizes the experience and reproduces it without going through the physical motions.

SUMMARY

Your capacity to perceive and interact with patterns of information can be affected by your vantage point. Most of us are used to having a limited vantage point. Because we assume a single, virtually two dimensional perspective is all we can perceive, we naturally limit our access to information. This is unnecessary. There are tools for accessing multiple vantage points and, thereby, facilitating your access to a broader landscape of information. The exercises described in this chapter are designed to: 1) provide you with multiple seats in the baseball stadium, 2) give you the 3-D perspective without your having to leave the comfort of your chair, and 3) let you access microscopic and telescopic views almost simultaneously. Much like George Carlin and the partially-filled glass, you can gain additional perspectives surrounding an issue not by changing the issue but rather by changing yourself.

Chapter 11

Mapping Information Terrains

For that one fraction of a second, you were open to options you had never considered. That is the exploration that awaits you. Not mapping stars and studying nebulae, but charting the unknown possibilities of existence. — Q to Picard, from *All Good Things*, Star Trek, The Next Generation[1]

CONCEPT MAPS

Concept maps are one of the basic educational tools. They allow you to depict how you visualize the landscape of your mind's conceptualization of a subject. Plus, they allow you to represent interconnections between patterns as you understand them. They provide a visual "frame of reference" and are simultaneously a knowledge base and an idea template.

Dr. Michael Zeilik of the University of New Mexico describes a concept map as "a diagram of nodes, each containing concept labels, which are linked together with directional lines, also labeled. The concept nodes are arranged in hierarchical levels that move from general to specific concepts."[2] The nodes are two dimensional and are usually represented as boxes or ovals. Dr. Zeilik adds that they "shift the emphasis from inert, static knowledge to contextually-embedded knowledge; from isolated facts to theoretical frameworks of related concepts."[2]

It is this aspect of shifting from isolated facts to theoretical frameworks that makes concept maps a key tool in the arsenal of achieving innovations and breakthroughs in research. In this chapter we will discuss key elements of concept maps, their historic use in education, their more recent application to research, and the opportunity to use dimensionality in

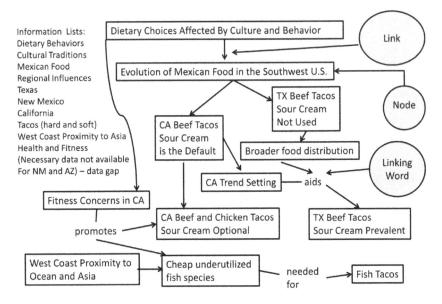

Figure 11.1   Basic Components of the Concept Map

working with concept maps as a means to access nonlinear knowledge better.

Figure 11.1 illustrates some basic components that go into a concept map. Information lists are often used as shown in the upper left. The first box, also called a "node," is usually the "proposition statement" or question (dietary choice and behaviors). Nodes are connected via links (uni- or bidirectional arrowed lines). Sometimes these lines are given labels (known as linking words). Subsequent boxes or nodes usually contain content as opposed to proposition statements (e.g., information about tacos in Texas). Data gaps can be identified as well. Here our lack of necessary information concerning the history of tacos in NM and AZ is specified in the information lists. Obviously, should this information become available, it might well cause us to modify this evolutionary concept map.

Once nodes are interlinked on a concept map, the overall appearance has some interesting implications. For example, concept maps represent a type of pattern much like an astronomical constellation. When the ancients viewed the visible stars, they envisioned linkages between them that formed specific patterns. From their frame of reference, these patterns represented figures that are now part of mythology: Orion, Sagittarius, Perseus, Capricorn, etc. The stars in a constellation provide minimal landmarks to enable the entire figure to be visualized. The spaces between the interlinked stars are filled in by the mind to create a solid figure. As an example, if Orion were not viewed as a human figure then the string of stars across the middle most likely would not be viewed as forming a belt.

Each constellation pattern became associated with certain characteristics in their relationships to other star patterns (location in the night sky, seasonality of appearances above the horizon, etc.). Naming a constellation was very much like naming and defining a concept map. It took on implications beyond a simple grouping of random stars. Awareness of the pattern and interactions with other patterns were also important. Applying these two aspects is important when constructing concept maps and using them in research as well as other areas of life.

Concept maps are not new to education nor are they new to education at Cornell University. In fact in researching materials for this chapter, I (RRD) discovered a rather embarrassing fact, a synchronicity. Concept maps were developed at Cornell University during the 1970s by a very distinguished then Professor of Education, Dr. Joseph D. Novak.[3,4] Dr. Novak credits their emergence from a foundation of earlier learning concepts discussed by David Ausubel.[5,6] The development and widespread educational applications of concept maps took place between the 1970s and 1990s at Cornell University at a time when entire curricula were being overhauled. Dr. Novak is presently an emeritus professor with Cornell and is also associated with the Institute of Human & Machine Cognition in Florida where his work continues. Dr. Novak and colleagues have published their work on concept maps in a series of books and articles.[7–9] Free concept mapping software is available at: http://cmap.ihmc.us

When I (RRD) decided to incorporate concept maps as a major component of *Science Sifting*, I researched their origins in detail. It was then that I realized that I had crossed paths with their inventor during the critical period when they were actually taking form. It is also the type of useful occurrence that begins to appear more and more often when we incorporate nonlinear aspects into otherwise purely linear scientific careers. The goals of enhancing the frequency of useful synchronicities and capturing and applying the full spectrum of available "ideas" to one's own research are at the very core of this book.

Back in the 1970s during my days as a newly-minted Cornell professor, I (RRD) had the opportunity to attend a new teachers' workshop taught in large part by Dr. Novak. That is where I learned what not to do in designing student exams. I also joined Dr. Novak around many a college committee meeting table learning from his significant experience at Cornell and discussing everything but his work or his then new book *A Theory of Education*[3] published by Cornell University Press. In those days, I was a scientific racehorse wearing blinders. I taught (avoiding exam design errors), met my committee assignments and otherwise focused exclusively on my own very narrow part of the research world. Educational theories were not my research and, hence, they were not my concern. So here I am now some 35 years later examining what had been happening a few feet away from that committee meeting table or teachers' workshop.

My first practical encounter with concept maps occurred back in the 1990s when they literally became the core basis for a complete overhaul of our curriculum in the College of Veterinary Medicine. Ironically, through a series of events involving closure of an academic department in 1991, I shifted from the college at Cornell (Agriculture and Life Sciences) where Dr. Novak was a faculty member to the college that was destined to become an experimental laboratory for his educational theories, the College of Veterinary Medicine. Spurred by the experience of Dr. Novak's protégé, Dr. Kathy Edmonston, and then Dean, Donald Smith, concept maps became the driving force for the complete overhaul of the DVM curriculum.[10] In fact, the restructuring of how we now teach veterinary students can itself be represented as a concept map as recently shown by Novak and Cañas in Figure 1 of their paper.[9]

Beyond showing our new curriculum as a pattern of interlinkages on one sheet of paper, concept maps permeate the nuts and bolts of our college curriculum. Our DVM degree students are taught to use concept maps to organize information and apply it to decision making involving real-life scenarios of animals presenting with symptoms of disease. Veterinary students build concept maps around fundamental science topics (e.g., different types of acquired immune responses) or around specific medical issues (e.g., urinary infections in dogs and possible

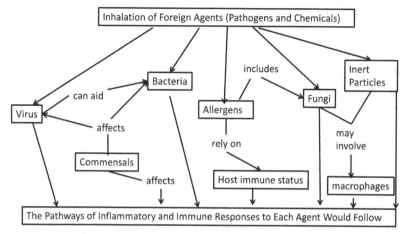

Figure 11.2   Beginning of a Concept Map on Host Inflammatory Responses

complications). I often ask my own veterinary students to build a concept map of inflammation since it is so central both to the effective resolution of infections and, when improperly controlled, to the promotion of most chronic diseases. If you can globally grasp all the various aspects of inflammation, that is an excellent entryway to engaging a majority of health issues. The beginning of a possible concept map on inflammation is shown in the Figure 11.2. Such a map is an entryway to examining all the components that go into host responses against exposure to agents (e.g., pathogens and chemicals). In the concept map, the respiratory system is depicted as a portal of exposure.

Figure 11.2 shows the first two sets of boxes that might appear in the beginning of a concept map dealing with inflammation. Each agent would elicit a different array of host responses. The concept map would provide a visual presentation illustrating not only all of the components involved in both acute and chronic inflammation but also similarities and differences in the responses produced against each agent.

So when your pet receives care from a Cornell-trained veterinarian, he or she is consciously or subconsciously working with concept maps to design the care that is most appropriate for your pet.

As diagrammatic representations, concept maps have several advantages. First, there is the opportunity to view the holistic realm of possibilities for your research topic. If, for example, you were working on a specific aspect of autism, you could include in your diagram all developmental, environmental, biological, chemical, and behavioral components of what is known or suspected about autism. Then you could view your narrow research aspect of it in the context of the broader array of possibilities. When this is used with the scientific method and hypothesis testing, there is an opportunity for constant adjustment and research questions to be asked within this framework. You may be asking a myriad of questions such as:

1. Who is at risk for autism?
2. At what stage of development are they most at risk?
3. What environmental factors contribute to risk of autism?
4. What physiological and biochemical changes promote autism
5. Does sex of the baby affect the priority of risk factors?
6. Is there more than one route to autism?

When one works from a template of information of the current understanding of basic developmental biology, physiology, genetics and then begins to ask these questions, concept maps can capture the interactions of patterns and their shifts as questions are asked and possibilities examined. Aspects of the concept map may shift as new information is introduced.

One example of the use of an open concept map is shown in Figure 11.3. In this case it is used to examine the comorbidities (associated secondary diseases and conditions that occur at an elevated rate) among patients with various primary conditions (*i.e.,* autoimmune thyroiditis, multiple sclerosis, psoriasis, inflammatory bowel disease, pediatric myalgic encephalomyelitis). These are also categorized as autoimmune or inflammatory diseases. However, they differ significantly in terms of the affected organs or tissues. This diagram shows the two secondary diseases/conditions that are common to all of these different primary diseases: depression and sleep disorders. This common association revealed in these concept maps helps to explain the increasing prevalence of depression and sleep problems among populations as well as the significantly reduced quality of life that is seen among patients with autoimmune/inflammatory diseases.[11]

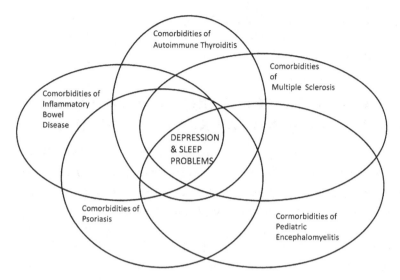

Figure 11.3 Free-Form Concept Map of Chronic Disease Comorbidities

Figure 11.3 shows an example of an open (also termed free-form) concept map showing a Venn diagram of comorbidities for several autoimmune and inflammatory diseases. Among the comorbidities that are common to all are major depression and sleep disorders. These are two conditions that one might not intuitively associate with these types of diseases. But we now know that both major depression and sleep problems can be linked to inflammatory dysfunction (information adapted from[11–13]). By using concept maps, these relationships appear and can serve as the basis for hypotheses and further testing. Unlike other types of concept maps, free-form maps are most helpful for revealing relationships but usually without an emphasis on hierarchical structure.[14]

As will be discussed in the following sections, the concept map has the potential for researchers to constantly focus in on narrow aspects of the idea network or in contrast to zoom out and test the validity of a narrow observation as being pivotal to the overall map. Additionally, we will discuss how concept maps are more than they initially seem in that they define a terrain with certain properties and significant possibilities. These can be termed "domains." Domains lead directly to the idea of three-dimensional concept maps. These terrains are also perfect for the use of fractals and fractal dimensions for this zooming in and zooming out viewfinder technique.

APPLICATION TO RESEARCH — A TEMPLATE FOR INNOVATION AND BREAKTHROUGHS?

Concept maps were originally formulated out of educational theory on how the mind learns and ways to maximize that learning. One could argue it is today's out-picturing of the historical axiom "a picture is worth thousand words." Although in the case of concept maps, they are not only a picture, they are also a process. In many ways, they are more like a thousand piece jigsaw puzzle. Once you fit the pieces together, you know how that terrain fits together. Given their origins in education, it is not surprising that most of the "proofs-of-concept" experiments about concept maps were closely tied with education and educational research rather than in other scholarly disciplines of study.

It should be noted that concept maps have the potential to take many different forms differing significantly in formality and approach. There is also some confusion in the literature concerning the definitions of concept maps vs. mind maps (Nesbit and Alexander, 2006, Novak and Cañas, 2008; Wheeler et al., 2009).[9,14,15] It is not the intention of this book to sort out these arguments and distinctions or to offer opinions of their merits. Instead, we will focus on the benefits and potential outcomes of applying concept maps to every aspect of research including not only specific research projects but also the career deliberations and decisions of the researchers themselves.

Among the first research efforts using concepts maps was that of Dr. Richard Iuli (Ph.D., Cornell University). He worked with both the Plant Pathology department at Cornell University and Dr. Novak to introduce the use of concept maps as an evaluation tool, both for individual research projects as well as multi-laboratory research programs. In a recent publication, Drs. Iuli and Hellden describe several important ideas about the utility of concept maps. First, they describe a process of collection of different concept maps from individual students followed by the development of shared understanding and a unified concept map from the group of students.[16] This is not unlike the integration that is discussed in the final section of this chapter on three-dimensional bubbles, except that there is unification across different types of concept maps rather than different versions of the same concept map.

The Cornell-Based Experiment

Dr. Iuli describes his experience with a multi-investigator, USDA-funded Rhizobotany Project involving several faculty members in the Cornell Department of Plant Pathology and the results as reported in his Cornell doctoral dissertation.[17] The goals of this application of concept maps to research were: (1) to help scientists work with their knowledge to better facilitate their research activities and (2) to help scientists enhance their understanding of the events and objects within their study.[16,17] The intriguing aspect of this application of concept maps in research was that the scientists were asked to focus both on their individual projects and on the overall group program project. Dr. Iuli found that some scientists were able to focus on their own project quite well and could seamlessly move to integrate a concept map of their own project with the group concept map that described the overall global Rhizobotany project. Other scientists used the narrow concept maps to a useful end but never were able to allow the individual and global concept maps to cross-fertilize each other.[16] In effect, these scientists were operating on a more limited playing field. Note that in describing his study, Dr. Iuli actually refers to the Rhizobotany Project Team research as "multidimensional" rather than using the term "multidisciplinary."

This is not to conclude that the narrow concept maps some scientific researchers use have no benefit. Evidence suggests they do, particularly when these maps are further engaged through the process of asking open-ended questions and subsequent revisions of the concept map are applied. But as Dr. Iuli has described it, these scientists who are working only within a single narrow concept map are working in "conceptual isolation." If given the choice, we would argue that conceptual isolation (which is not the same thing as increased focus of attention) does not generally enhance one's likelihood of innovation and breakthroughs in research.

In contrast, in the examples where team scientists could work between more than one research concept map (in this case, local and global), they appeared to be better able to refine and work with the knowledge contained in both maps. This aspect of multidimensionality will be discussed straight-on in the final section of this chapter as we believe this is one of the most useful tools that a researcher can learn.

In the same paper, Dr. Iuli states that "it is essential for scientists to have an understanding of the nature of knowledge and scientific inquiry if they are to be exemplary researchers and teachers."[16] He goes on to say that "only then will they have the creative insight to make quantum leaps in our understanding of objects and events in the natural world."[16] In fact Dr. Iuli points out that the research career of Cornellian and Nobel laureate Dr. Barbara McClintock demonstrated that "personal convictions on the nature of knowledge extend beyond the cognitive."[16]

The Use Of Open-Ended Questions

Wheeldon and Faubert[14] discuss several benefits of applying concept maps and the exercise of preparing them for broader research applications and, in particular, quantitative research. Some differences were noted based on whether formal or informal processes were used in the preparation of the maps (e.g., those generated using computers vs. those drawn free hand). There were also certain specific common benefits that spanned the process of map generation. Importantly, these researchers argue that concept maps provide a "framing experience."

The beginning of "the framing experience" seems to be much like positioning participants at a series of doors with each representing an entranceway to a space containing additional insights. The potential is now in place for a dramatic expansion of one's perceptional horizons. What Wheeldon and Faubert[14] did next appeared to allow the concept map preparers to intensely frame the experience, to select a door and then begin to step through the door into a much-expanded terrain. What initiated the experience was a matter of them asking a series of common, open-ended questions concerning their maps. The role of the open-ended questions cannot be overemphasized. In many ways it is the key in the lock once the concept map has been prepared and they are positioned to move through doors to access additional information. The open-ended questions asked by the researchers overlapped greatly among the participants and differed only slightly based on the formality of the maps each had prepared.

Concept maps not only ground participants but, when participants frame their experiences, the framework also provides doorways for unique types of experiential recall. This process is useful and novel in that it, as discussed by Legard[18] and Wheeldon,[19] can help to unlock unique memories. Wheeldon and Faubert[14] argue that it allows participants to "consider a prior experience in a new environment" or "think about prior experiences in different ways."[14] These observations of framing during the process of making concept maps and engaging open-ended questions are pivotal for a type of scientific breakthrough. These breakthroughs are driven by the interlinkage between patterns of seemingly disparate information. The question is: what happens when you unlock unique memories that are also linked with huge patterns of additional information and do so in the context of current research problems? Is that akin to Nikola Tesla using Goethe's words in Faust to discover modern electromagnetic theory (see Chapter 12) or Steve Jobs using his old calligraphy course information to create Fonts (as discussed in Chapter 2)?

This process as perceived in the present discussion could be diagrammed as shown in Figure 11.4.

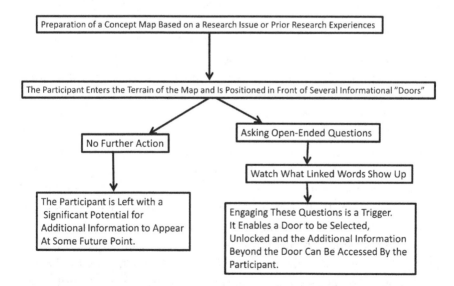

Figure 11.4 Gaining New Research Insights by Using Concept Maps

Figure 11.4 shows the stepwise process for using concept maps to facilitate the likelihood of major research insights and possible break-throughs. 1) It begins with having participants create a concept map. This places participants within a unique landscape of organized information that is directly linked with the participant's experiences. It also has the effect of placing the participant at a series of new doorways, which are potentials for accessing and bringing in additional information that is lying in a terrain just beyond. 2) A door is selected and unlocked, and the participant moves through the doorways to the new information by the process of engaging a series of open-ended questions surrounding the original version of the concept map. In many cases, the new informa-tion may be within the participant's realms of prior experiences. However, the connection to that information and the current information embedded in the concept map has yet to be made. This two-step process appears to enable multiple patterns of information to be combined, which produces a totally new set of perceptions (a revised concept map).

Domains

Once a concept map has been developed, there is the formation of a ter-rain. These are areas of information that do not exist until concept map boxes and interlinkages have been defined. Essentially, this is the area in between points on a grid. When we form grids or concept maps, we usu-ally emphasize the importance of the points on the grid (or boxes on a concept map). But the spaces in between the points are equally important when it comes to information. For our purposes, these spaces on a grid will be termed "domains."

An example of such terrains is in Figure 11.5. In this figure, four concept map boxes are formed around the topic of whether a favorite U.S. National Football League (NFL) team will make the playoffs in the present year. While this might seem to be quite a simple issue, it is not. Several factors go into this issue. It all starts with the season schedule, which is usually released by the NFL several months in advance (while certain games are completely prescribed by the team's league and division affiliations, more than half the games are uncertain until the schedule is announced). The first level of outcome pertains to the number of games my (RRD's) favorite team

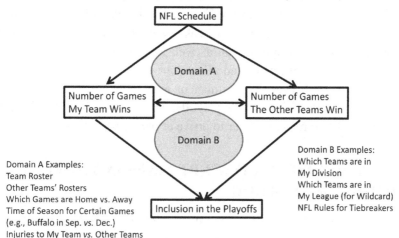

The Chance My Favorite NFL Team Will Make the Playoffs This Year

Figure 11.5 Concept Map Showing Domains

wins. Because they are playing other teams in the league and division, my team's number of wins affects the number of wins by my team's competitors. If my team wins, someone else's team loses. This relationship of schedule to wins by my team vs. the other teams can be depicted as a triangle. The existence of that triangle now defines a terrain that did not previously exist. This is labeled in Figure 11.5 as Domain A.

Domain A contains information that may be as important as the information in my concept map boxes. Some of the information may be known when the concept map is originally constructed while additional information may appear as research progresses and the concept map evolves. The dynamic concerning team wins involves many factors such as comparative rosters (with opponents), which games are at home (vs. away), playing in adverse weather (e.g., a warm weather-based team playing in Buffalo, NY in the winter) and, of course, key injuries (e.g., losing your star quarterback for the season with an injury in the first game). Other variables may not be obvious but will dwell within the now-defined Domain A terrain.

By establishing the relationships among the four boxes in Figure 11.5, we also identify a second terrain. This is the area defined by the bottom triangle. It is labeled as Domain B. Like Domain A, Domain B has both immediately obvious information as well as the potential for additional information that connects to and affects our issue (making the playoffs). Domain B exists because the number of wins alone does not determine whether my team will make the playoffs. There are contextual comparisons that must be made in order to arrive at the answer.

USING MULTIPLE CONCEPT MAPS IN THREE DIMENSIONAL SPACE

Concept maps are normally a two dimensional representation of your best understanding of the world surrounding a given topic (*e.g.*, potential pathways of an allergic response in the lung) or platform (*e.g.*, the cardiovascular system, a cytokine network). We have already discussed how the preparation and engagement of a concept map opens up the potential for additional research insights and possible breakthroughs particularly via the use of open-ended questions. As a two dimensional representation of information, can concepts maps be raised to a higher order map for research effort and intellectual engagement? Is there a way to create a three-dimensional construct? One way in which this seems to occur is through nested matrices. If a concept map is viewed not simply as a grouping of information on a flat sheet of paper but instead as a free-floating form within a bubble, then there are additional informational interactions that can occur outside of two-dimensionality. Multiple concept maps can be added to the 3-D bubble as if each one were like a pane of glass. But information flows not only within a single concept map but also between concept maps with potential relatedness. Each map can be stacked like panes of glass within the 3-D bubble, and one can examine the interactions of information among the different concept maps (plus the acquisition of additional information) within a three dimensional bubble. It is basically the intercalation of information that can occur in this space. Plus, there is the opportunity for both linear and nonlinear results to occur in working across concept maps.

As an example, we will describe in some detail how we used this three-dimensional bubble idea in a project that has grown exponentially to encompass history, culture, decorative arts, and social sciences. This project began simply enough. It started as a straightforward examination of what began as a hobby and type of investment, antique Scottish silver (a rather narrow topic within the decorative arts). It rapidly became much more than what it originally seemed. We wanted to build a map around:

1. What silver had been made by Scottish goldsmiths as per forms (*e.g.,* teapots, sugar bowls, creamers mugs, salvers, candlesticks, coffeepots) and types (particularly forms unique to Scotland (*e.g.,* thistle mugs, quaichs),
2. What seemed to be extant today (*e.g.,* surviving from 16th, 17th, 18th and 19th centuries),
3. What was comparatively rare (*e.g.,* Scottish-made dish crosses are rare, English ones are not).

The self-serving purpose was that as academicians, if we were going to buy any antique silver, we wanted to be informed, buy wisely and avoid mistakes. At least that is what we thought we were doing. This initial mapping goal became a 6,000 entry organization (by form, chronology, maker, description) of all known antique Scottish silver at the time of publication. In the absence of other mapping efforts, it is a potentially useful reference.[20,21] So that is the first concept map for our bubble. An example of the start of this concept map is shown as follows with the only information thread shown being that for teapots.

Figure 11.6 shows the beginnings of a concept map using the 6,000 entries of Scottish silver from editions of the *Compendium of Scottish Silver.*[20,21] Because each entry contains information on the form of the item (e.g., teapot), the style (e.g., a spherical-shaped teapot with a straight spout), the dimensions, the decorations, the goldsmith who made the item (e.g., James Ker), the town in which it was made (e.g., Edinburgh) and year of production (e.g., 1735–36), it is possible to track the emergence of forms and styles and learn which makers were most prolific and which first created each new design. If this concept map is paired with additional ones (such as one describing the training of goldsmiths), it is then possible to see

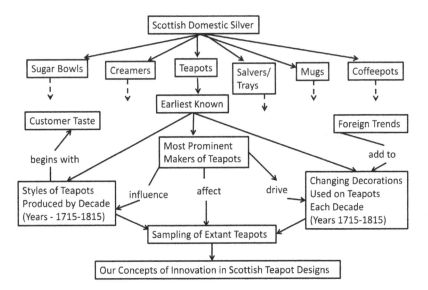

Figure 11.6 Beginning of a Concept Map on Extant Antique Scottish Silver

how designs progressed and evolved as specific masters, in turn, taught their apprentices their craft.

As just mentioned, the second map came about as an effort to map the training of the approximately 1400 individuals including approximately 420 eventual freemen goldsmiths of Edinburgh (the premier center for Scottish goldsmithing) between the 1500s to the 20th century. This resulted in a series of training trees that included the 1400 entries (actually over 110 individual yet interlinked maps) that also included demographic information.[22]

The third concept map for our bubble was a genealogical survey of the same 420 freeman goldsmiths mapping their ancestry.[23] Finally, a fourth map was created utilizing significant events in Scottish history over the same 500 years that affected the structure of society. Figure 11.7 shows these four maps as two-dimensional structures that have been introduced into a three-dimensional playing field (the bubble). They are essentially a series of planes (nested matrices) that can now interact. The effect of allowing these maps to interact (i.e. interactions among patterns of information) was striking.

An Example of Using Multiple Concept Maps in a 3-Dimensional Bubble

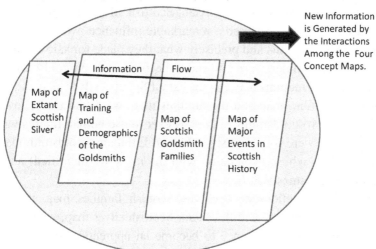

Figure 11.7   An Example of Using Multiple Concept Maps in a 3-Dimensional Bubble

Figure 11.7 shows four separate concept maps that concern some inter-related topics. Each is a two-dimensional sheet of paper or pane of glass. They have been accumulated in a sphere or bubble that is three dimensional in space. In this space, information in the four patterns represented by the four concept maps can interact. This produces sometimes quite surprising insights that could not be accessed by working only with a single concept map.

Here is a sample of the information that was previously unknown but showed up only after these multiple concept maps were allowed to interact in a 3-D bubble.

1. Through interactions between the History of Scotland concept map and the Extant Scottish Silver concept map, we found the following: The Union of Scotland with England (1707) and the Battle of Culloden (defeat of the Jacobites) (1746) were the two most significant events that resulted in a loss of Scottish identity as reflected in both Scottish linguistic and decorative arts forms.[24]

2. Through interactions between the Scottish families map, the gold-smith's training map and the extant Scottish silver map, we found the following: Women exerted a remarkable influence over who became Scottish goldsmiths and precisely what they made for sale despite the fact that no women were allowed to become freemen goldsmiths.[25]

3. Through interactions between the extant Scottish silver map, the gold-smith's training map and the families map, we found the following: The only two goldsmiths in the UK ever to use human figures within their mark were a previously unknown Edinburgh goldsmith (James Mitchell — whose association with the mark we identified) and his apprentice, James Welsh.[26]

4. Through interactions between the Scottish families map, the gold-smith's training map and the extant Scottish silver map, we found the following: The best chance to become an apprentice to a goldsmith (immediately after father-eldest son relationships) was to have a mother, sister, or aunt who had a maternal family connection to the goldsmiths. Surprisingly, this was more important than having a strong business connection.[23,25]

Importantly, much of this type of information is quite nonlinear. It emerged only after multiple concept maps were allowed to interact. This is one of the potential benefits of working with multiple maps in 3-D space.

Obviously, the prior example is quite specific and pertained to a subject outside of the natural sciences. But the same strategies can be useful across the areas of science and technology. For example, theoretical concept maps can be generated surrounding particle physics. This could include: 1) research facilities, 2) funding, 3) the prior history of particle discovery, and 4) anticipated areas of future research. An example of this type of map interaction in a 3-D bubble is shown in Figure 11.8.

Figure 11.8 illustrates a theoretical example following the same model as the already-used example on Scottish silver and history shown in Figure 11.7. In the present case, theoretical particle physics is the topic. Interactions among the information contained in four somewhat related, two-dimensional concept maps will occur in this 3-D bubble. This provides a potentially useful new terrain through which new insights can emerge.

An Example of Using a 3-Dimensional Bubble Involving Particle Physics

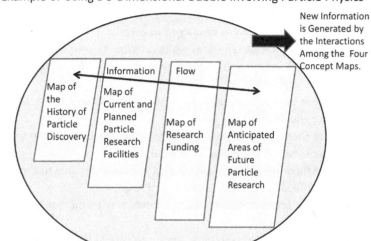

Figure 11.8   An Example of Using a 3-Dimensional Bubble Involving Particle Physics

CREATING CONCEPT MAPS

The following exercises are designed to allow you to try your hand at making concept maps. First, you will create a simple map as a foundation. Then you will modify it to open up the terrain. A second concept map will provide some relatedness. And finally, your two maps will interact in three dimensional space. The topics used in these maps will go to the heart of the goals of *Science Sifting*, which is to prepare students interested in research and researchers in training for a productive and creatively innovative career.

CONCEPT MAP EXERCISE 1

1. Consider the next five years of your educational and research career.
2. Design a concept map detailing the preparations to do research that you expect to undergo.

CONCEPT MAP EXERCISE 2

1. Take the concept map you developed in exercise 1.
2. Answer the following questions while referring to your map.
    a. Are there any training experiences that could be useful but might seem impractical?
    b. If you look at your career in 10 years, would you add any additional preparation now?
    c. Are there minor experiences in your past that could help to propel you forward? Have you considered those?
    d. Are there any new factors or events you could envision that might add additional branches to your concept map?
    e. If your research career changes focus, how might that look in your concept map?
3. In light of your answers to those questions, would you modify your concept map in any way? If the answer is yes, do so now.

CONCEPT MAP EXERCISE 3

1. Design a parallel concept map detailing your personal life goals over the next five years.

CONCEPT MAP EXERCISE 4

1. Take your amended first concept map about your research preparation and your second concept map about your life goals.
2. Punch a hole in the upper middle of each.
3. Slip either a key ring or a piece of thread through the holes to tie the two maps together. The maps are now interactively joined in a three dimensional space.
4. Looking at your maps from this perspective, would you now change any aspect of either concept map?

Do you have any impressions surrounding these topics even if they fall outside the two concept maps?

The purpose for linking these exercises is not to provide instant new insights concerning your career and personal life, though kudos if this happens. The purpose of these exercises is to give you a frame of reference for engaging multiple patterns of information in such a way as to increase the probability of having breakthrough insights. The more patterns of relatedness you can engage, the more likely nonlinear concepts will emerge from the interactions of those patterns. Engaging in these exercises is more important than any immediate quantitative assessment of their effectiveness.

SUMMARY

Concept maps may be adapted for use in any research setting (not only their original use in the social sciences). Because they define a landscape or terrain, they provide a two-dimensional template for visually observing shifts as the researcher acquires new insights. Node points on a concept map can define the "space in between," which are termed here as "Domains." These domains are rich sources of additional information and should be viewed as such.

One of the ways to encourage shifts beyond the incorporation of purely linear information is to ask open-ended questions and re-engage the concept map during the process of responding to the open-ended questions. A second way to increase the level and type of information available to the researcher is to work with multiple concept maps in relationship to each other. This is what we have termed working with maps in three dimensional space. This offers the possibility that information will appear that is beyond the sum of the parts of various concept maps. Examples of this are shown from our own research into the history of Scotland and its cultural and artistic traditions of goldsmithing.

Chapter 12
Pattern Jumping and Lateral Thinking

Nature uses only the longest threads to weave her patterns, so that each small piece of her fabric reveals the organization of the entire tapestry. — Richard Feynman[1]

There are only patterns, patterns on top of patterns, patterns that affect other patterns. Patterns hidden by patterns. Patterns within patterns. — Chuck Palahniuk[2]

The way is long if one follows precepts, but short if one follows patterns. — Lucius Annaeus Seneca[3]

Pattern Jumping is the name we have given to a process that facilitates access to road-blocked patterns of information by sifting for depth of information within unrelated patterns. You work where you can and then return to the roadblock pattern having newly-gained potential. Processes that are similar to pattern jumping have been known by other names such as lateral thinking as described by numerous authors.[4–6]

Lateral thinking has been described as pursuing problem solving by using unorthodox or apparently illogical means. Only, you don't recognize the means as being illogical until after the entire process has been revealed and the nature of the roadblock has been discerned.[5] According to de Bono,[7] continuing to dig your hole deeper in the same place is not the same as beginning a new hole in a different place. In other words, if you aren't able to see the whole pattern then your vantage points are blocked or ineffective, and you have no access to the patterns holding solutions.

Linear or vertical thinking, which we view as continuing to ram your head into the wall: 1) is going to take time (which you may not have),

2) is likely to come at a cost (your head), and 3) still may or may not work. The importance of being able to shift out of vertical thinking into lateral thinking is reflected in the fact that Mayo Clinic educators recently argued that a continuing focus on vertical thinking impedes the effective education of physicians.[8]

For our own applications of using unrelated patterns to access the one of primary concern, we prefer the term pattern jumping (PJ) to lateral thinking. Part of the reason is that we are not totally convinced how much formal "thinking" is involved. PJ best describes what we have experienced and defined as jumping with ease from one large organizational structure of information into a second equally massive but completely unrelated grouping of information. Additionally, PJ extends beyond preformed lateral thinking puzzles. With PJ you are usually constructing the parallel puzzle yourself. But like lateral thinking puzzles, you do use the momentum acquired on one pattern puzzle to gain better access and/or more depth and springboard ahead on your second pattern puzzle.

If you want a visual example of PJ, you might think of it as playing two side-by-side games. One is a 3-D chess game played against a similarly-skilled opponent, the other is a game of chutes and ladders played with your 8 year child. After a few moves against your opponent on one board, you move to the alternate board and continue that game. The games are very different as are the opponents. Despite these differences, the progress you achieve in one game is fully capable of bleeding over to the second game. Each time you switch between games, you take new insights with you. If this seems illogical, that is fine. The progress with PJ is not linear.

In fact, PJ relies on what Daniel Pink[9] in his book, *A Whole New Mind*, has categorized as "Symphony." Pink defines symphony as "the ability to put together the pieces...the capacity to synthesize rather than analyze; to see relationships between seemingly unrelated fields; to detect broad patterns rather than to deliver specific answers; and to invent something new by combining elements nobody else thought to pair."[9] When you engage one pattern that is available to you and then use this to detect and reveal the previously unavailable pattern of your research issue, you create the symphony.

USING PUZZLES: NUMBER, WORD, SPATIAL-ORGANIZATION

PJ probably occurs when individuals use word or number puzzles as a pattern-based diversion prior to jumping back to their research roadblock. Crossword puzzles are considered among the ultimate left brain-dominant games. But keep in mind that while you engage the information as one or two words at a time, the overall information is laid out in a specific pattern and many cross words have overarching themes. Therefore, they are the perfect tool to intermingle if not force your left brain into integrative, nonlinear strategies for overcoming roadblocks. Think of it as a type of Sherlock Holms meets Salvador Dali or Jimmy Hendrix exercise. In fact, there are benefits to be derived from taking an activity that conventional wisdom might label as narrow and showing that it can serve a much larger purpose. Many consider crossword puzzles to be a pastime activity that is useful for sharpening memory and word usage or delaying age- or condition-associated memory decline. Studies have suggested that crossword puzzle participation among older individuals with dementia has delayed the onset of memory decline.[10] When you delve deeper into crossword puzzle patterns and then transfer that momentum to other issues on a regular basis, you are training to become a type of MacGyver (a US 1980s-era TV show and character) of the research world.

Ironically, crossword puzzles have increasingly found their way into biomedical education as effective tools for learning.[11] Educators at the University of Saskatchewan Medical School found that students not only performed well using puzzle-designed, course material, but also took it upon themselves to transfer this learning tool for use in other courses.[12] Use of crossword puzzles to enhance biomedical education is becoming a worldwide trend.[13,14]

Another variation on the crossword puzzle is Sudoku. These number puzzles represent a special form of mathematical patterns. The form they take is that of 9X9 grids with 3X3 subgrids. They represent a type of matrix. Unlike crossword puzzles, with Sudoku a partially completed grid is provided to the puzzle player, and he/she must see and then complete the numerical pattern. Variants exist in which alternate sized grids are used (e.g., 4X4, 5X5, 7X7, 16X16) or letters are employed instead of numbers (a variant called Wordoku). As with crossword puzzles, working with Sudoku puzzles has been reported to benefit patients with neurodegenerative conditions. Nombella *et al.*[15] reported that training programs with Sudoku puzzles provided patients with Parkinson's disease significant improvements in cognitive reaction times and correctness of answers. Additionally, based on MRI evaluations, Sudoku-trained patients had cortical activation patterns that were more similar to those seen with controls than occurred with untrained Parkinson's disease patients.[15]

The Sudoku square has a particular mathematics surrounding it. It is a type of Latin Square, which is defined as an $n \times n$ array filled with n different symbols, each occurring exactly once in each row and exactly once in each column. But Sudoku goes further by using the restriction that each number appears only once within the nine squares of each 3X3 block. Restrictions within Sudoku puzzles have other effects. Because randomly-generated numbers could have a number appearing more than once in the grid, the numbers in 9X9 Sudoku puzzles are reported to have a greater Shannon entropy, a quantification of the expected value of information present in a message.[16]

Other types of spacial-organizational puzzles have been used to engage and open up patterns. Jigsaw puzzles and Rubic Cube-type 3-D puzzles can serve as entryways to creative, organizational-level problem solving. In fact, in his business creativity workshops for groups, Robert Lucas[17] uses jigsaw puzzles as an opening exercise to prepare participants for engaging the more complex patterns within a business. Working on one pattern can pave the way to progress on another where you either cannot see an opening or are stuck.

In an intriguing analogy to jigsaw puzzles, Robert Weisberg[18] describes the work of Watson and Crick in elucidating the structure of DNA. Of interest is the fact that Watson and Crick were originally working on models of 3D structures that were not a helix-shape. They hit a roadblock. However, they then took a different approach by setting a helix shape as a default assumption and looking at how the data pieces might fit that 3-D model. To Weisberg this is like first working on a jigsaw puzzle where you have no final image to work toward vs. seeing a final picture (the helix) and then asking how the puzzle pieces might fit an overall template.

USING DIVERGENT SUBJECTS TO GAIN INSIGHTS

PJ may occur between such totally different subjects that you could well be the only person on the face of the earth to ever engage both of these topics during the course of single day (e.g., changing global distribution of *Barbula indica* moss the past century as one pattern and the distribution of opera aria reuse by Rossini factored by his age vs. similar reuse by his contemporaries). The goal is for a researcher to engage one type of complex pattern of interest and then, with intention and recognition, leap over to the organizational information that is part of researcher's current roadblock. One example of this can be found among the interests of Sir Peter Medawar who was already mentioned. The transplantation immunologist and avid opera lover found that the best way he could envision how lymphocyte trafficking occurred was to describe it in terms of a local production of the Gound opera, *Faust*. He observed that the lymphocytes are like the "chorus of soldiers in a provincial production of Faust – they make their brief public appearance and then disappear behind the scene only to re-enter by the same route."[19] To Medawar, the patterns within the immune system were not unlike patterns he knew in operatic scenes. He could jump between such patterns with ease.

Certainly, the insights that arise from pattern jumping can happen to researchers by sheer accident. But they can also be facilitated with purpose. Medawar understood his own capacity to delve deep into opera and then into the inner workings of the immune system. Once we lay the groundwork, we will describe how PJ can be used purposefully to increase the likelihood of overcoming roadblocks.

Pattern jumping is probably not solely a left-brain or right-brain function. It is likely to require hemispheric cooperation and high-level executive brain function use. Working with the details of a pattern, some level of special arrangement and drilling down to deeper levels would likely be considered as left brain activities. However, working holographically and zooming out to capture the pictorial impressions of the pattern as you work would likely engage the right brain. The jumping or transference itself is certainly nonlinear since it involves going from one holograph to another.

But that is not to say it does not have left brain cooperation. In fact the whole exercise is probably designed to promote better hemispheric cooperation. However, until you actually experience and recognize PJ, it may well be blocked by The Three Blind Mice. You may have no frame of reference for being able to shift among large unrelated patterns and have your progress in one pattern translate directly to progress in the other. This is an experience where noticing it happen just once can provide you with the opening to pursue PJ as a useful strategy to enhance creativity.

NIKOLA TESLA AND PJ

One of the most famous examples of PJ concerns Nikola Tesla and one of the greatest scientific discoveries ever made; his discovery of rotating magnetic fields and the foundation stone of the alternating-current motor. In his autobiography,[20] Tesla describes how this monumental discovery happened.

It was February in Budapest where Tesla was recovering from a serious illness. He was able to take a walk with his assistant in a park along the Danube River as the sun was beginning to set. Tesla had a great love for the works of Johann Wolfgang von Goethe, in particular *Faust*. While watching the sun's rays Tesla began to recite Faust's monologue from memory in which the character is speaking of the sun's fading rays and with his disillusionment with a life almost totally devoted to science and knowledge. (Note, there is considerable irony in this being the selection since Goethe himself had considerable physics education and had written a monograph on light, color and optics.) Tesla reported that he noticed a sort of flash of light and then used a stick to begin drawing in the sand what became the basis for the AC motor. His assistant immediately recognized the significance of the drawing.[20] A chart depicting the relationships between Tesla, Goethe, and Faust is shown in Figure 12.1. Tesla was able to dive deeply into a pattern involving the polymath Goethe and his work about a frustrated scientist and then "jump" back into the depths of his own road-blocked, research pattern emerging with a nonlinearly-derived solution.

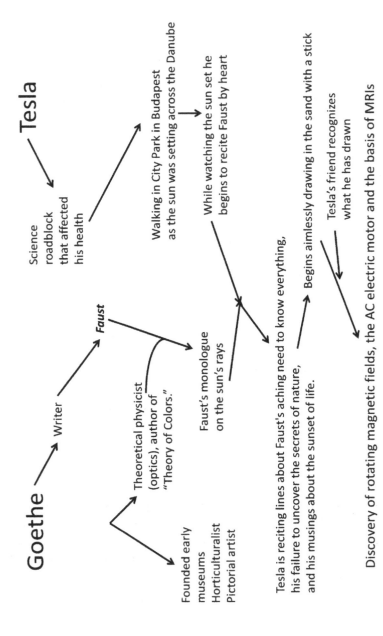

Figure 12.1   Pattern Jumping — Nikola Tesla

A PATTERN JUMPING PROTOCOL

The following provides an example of a stepwise approach for PJ.

1) You can take any activity where you have a real interest (e.g., a pastime, hobby, other scholarly interest, family activity, home improvement project) and start to recognize the patterns you are encountering with this non-work activity.
2) Once you are used to engaging the non-work pattern, focus on delving deeper into it and watching it change. This does not have to happen all at once. You can return to it periodically to work further on perceiving more of that non-work-related pattern.
3) Then jump back to the issue where you hoped to make progress at work and notice where new information is now available. That is your access to the previously road-blocked pattern.
4) If need be, repeat this process by going back to the leisure pattern and working further on it, then returning to the work-related pattern yet again. No leisure interest is too trivial. You can jump between them multiple times to greater benefit.

The act of drilling deeper and expanding your view of the non-work-related pattern seems to open the access to the target-problem at work.

Originally, we had simply noticed this outcome as an unexpected added bonus following the pursuit of leisure interests. But it did not take many examples of this pattern jumping as a useful outcome until we started to use this approach as a purposeful strategy to make progress at work. Of course, have fun telling your department chair you have to go play a round of golf so you can formulate a next generation scientific hypothesis or design your next revolutionary experiment.

As a side note golf was only an example. That is not a jumpable pattern for me (RRD) personally. I was once given sound advice by an acquaintance to the effect that if I enjoyed having friends, I should never play golf with them. It was a testimony to my lack of prowess for the game and defined what I have since come to call the Grass Ceiling.

PERSONAL EXAMPLE OF USING PJ

To give you a clearer picture of exactly how PJ has worked for us, we will describe a very specific example: pattern jumping between one pattern, historic patterns of Scottish decorative arts, and a second pattern, today's patterns of comorbid chronic diseases. These are unrelated patterns, and you would not expect to find the range of material in any single journal, classroom lecture or book of any kind (until this book). They do not routinely go together like ice cream and cone.

As discussed in a prior chapter, a hobby area of interest for us is Scottish silver. This includes: (1) its history since the 1500s, (2) the craftsmen who produced the gold and silver, (3) their training, (4) their demographics and how they were chosen as apprentices, (5) their output, (6) the creative flow of designs over the centuries, (7) the customers who supported the craftsmen and their changing tastes, (8) the political and religious connections, (9) the allegiances among the goldsmiths and, (10) their family associations. It is a serious hobby in that it has produced five books and four scholarly articles covering these specific topics

One of the approaches used to investigate the 1400 apprentices, journeymen, and master goldsmiths of Edinburgh for an approximately 500 year span was to diagram each component of interest. For example, one component was their training. Remarkably, some of the master-apprentice training pedigrees were unbroken for 15 generations (almost 500 years) and led us to a wonderful face-to-face meeting with the last surviving master of that continuous training program. But diagrams of output, political events (*e.g.,* the Battle of Culloden, The Union with England) were informative as well.

Finally, at the suggestion of George Dalgleish, Silver Curator of the National Museums of Scotland, we examined the family connections. Here, a completely unexpected pattern emerged. Direct father-son training of goldsmiths was expected and easily seen based on shared surnames. But that explained no more than half of all of Edinburgh's goldsmiths. Where did the remainder come from, and how did they get their chance at this highly desired profession? The answer is shown in the figure that follows, which is based on our published findings. More than half of the Edinburgh goldsmiths depicted in this pedigree tree were also relatives but came to the profession via matriarchal connections (mothers, sisters, cousins, aunts). This role of women in determining who got to be a goldsmith had been completely unexpected. It explains how David Mitchell, Thomas Mitchell, Robert Inglis and Edward Lothian all became highly prolific and important goldsmiths. These and related findings led to an article on the improbable subject of the importance of women in Scottish goldsmithing published in *History Scotland Magazine* in 2011.[21] The subject was improbable since for hundreds of years, there were no women goldsmiths save a handful of widows permitted to continue the family business. But that was only a superficial observation. Much more lay beneath the surface as shown in Figure 12.2.

You may be questioning why you are reading such details on a topic quite unrelated to science and, frankly, a bit obscure. The reason is what happened next. Around the same time we were playing with overlaying Scottish history patterns, I (RRD) became interested in why so many seemingly unrelated chronic diseases with early-life origins (e.g., childhood asthma, type 1 diabetes) were increasing in prevalence. One idea was that these diseases and conditions were not as unrelated as they seemed on the surface. Perhaps there was an unseen pattern to be revealed. So I began jumping back and forth between working on the goldsmith patterns and patterns of chronic diseases. With the help of several scientific collaborators, we produced a series of papers on immune dysfunction-inflammation-related chronic diseases. The last showed a pattern diagram not unlike that used in the goldsmiths' books. But in this case what became clear was that these diseases occurred in different tissues and organs, were treated by different medical specialists and were medically coded as having different origins but were not that different. They had overlapping root causes as well as comorbid health risks. These findings were published in the journal of the NIEHS, *Environmental Health Perspectives*[22] and helped to explain not only the probable sources of conditions like depression, sleep problems and sensory loss but also a major route to target tissue-specific cancer.[23]

Without working back and forth between science and history and overlaying the Scottish history patterns, I would never have seen the overlays of the chronic disease patterns. It only came into my awareness after I was deep into the history overlays. That is the essence of PJ. Immerse yourself in a pattern of interest that is open to you and then use that immersion as access to move into the pattern that surrounds your road-blocked scientific issue.

Here is a way to ease into PJ:

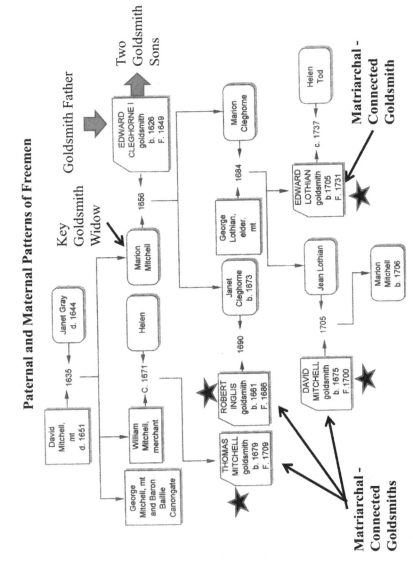

Figure 12.2 Edinburgh Goldsmith Families *c.* 1650–1750

THE PJ PAPER EXERCISE

1. Take your current most significant dilemma at work, and write a brief description of that in the center of a piece of paper.
2. On a second sheet of paper, take a topic you (1) enjoy and (2) know something about. Write that in the center of the paper. The possibilities are endless but here are examples:

   a. your favorite sports team (at any level and, yes, curling is a sport)
   b. the protagonist in your favorite novel, play or movie
   c. your community's K-12 educational program
   d. photography
   e. fishing
   f. your favorite video game
   g. songs or song artists
   h. family history
   i. amusement parks
   j. hiking trails
   k. spa treatments
   l. martial arts
   m. gym usage
   n. holiday shopping
   o. cooking
   p. local history

3. On the second sheet, begin to diagram the positive and negative factors that you perceive affects the topic of interest.

   For example: a) you might work on the potential success of your favorite sports team in the upcoming season such as new players, retiring player, injuries, coaching changes, difficult travel schedule; b) you might diagram a protagonist's goals, talent's, limitations, tendencies, supporters, antagonists, wildcard factors; c) you might examine Black Friday shopping strategies (US tradition) such as, which stores to visit in which sequence looking for which priority categories and including all the pitfalls that need to be considered.

   You can formulate the diagram any way it shows up for you. Just be sure to connect the pieces into some type of pattern. Notice both the individual pieces you add and also the overall pattern that is formed.

4. Now return to the first sheet of paper with your work issue. Is there any factor you want to add to the issue's name? Place it anywhere it shows up connected in any way that makes sense to you. If nothing shows up, no problem.

(Continued)

THE PJ PAPER EXERCISE (*Continued*)

5. Return to the interests sheet and examine it further. How would you modify it now? Are there more factors to add? Would you rearrange anything? Would you change the relative size of factors based on relative importance? What additional insights do you have upon this second glance?
6. Once again go back to the work issue sheet. What do you want to add here? What can you see now that was not apparent when you previously looked at the sheet?
7. Not surprisingly, you can repeat this process yet again and do so after taking a break (e.g., a few hours or a day later).

This back and forth process using both the drilling down of added pieces to the pattern and also standing back to see the pattern's overall terrain is very powerful.

There is another useful benefit of PJ. In a prior chapter, we discussed the unhelpful mantra that says you need to work harder to see more results. Most researchers have had this concept engrained into them. We are not encouraging a lack of productivity, but we are suggesting that working longer hours is not the only path to success. When you start to use processes like PJ, your leisure time can become just as useful for your work as the time you actually spend at work. With PJ, the fabric of conventional wisdom begins to fray a bit if not completely unravel.

SUMMARY

Pattern jumping (PJ) enables you to use the progress you have made in non-work-related activities to facilitate your research breakthroughs. By delving deeper into complex patterns unrelated to your work, you can gain better access to the work-related patterns where you have been road blocked. The process is not linear and, as a result, you may find some personal resistance to this entire approach. But once you experience PJ you will have a calibration for its use. You will also realize that such a strategy can be applied with intent as a highly-useful tool for multi-tasking. If you remember that the work-related problem and the favorite hobby subject are both just patterns of information, there is no reason beyond your own self-imposed rule set why PJ cannot be applied.

Chapter 13

Synchronicity

It's déjà vu all over again. — Yogi Berra[1]

We have already discussed pattern jumping and presented examples. But these examples pertained to jumping between two patterns of your making, although many other people are likely to share your research area or your hobby interest.

One of the questions now might be, "Can you access patterns of information that were already engaged by others?" In other words can you tap into preformed patterns and patterns that extend beyond yourself to include others. This is useful on several fronts. For example, can you get the same vantage point as a reviewer on one of your papers? Is that necessarily different than shifting seats in the baseball stadium?

For this chapter on synchronicity and collective consciousness, we enter the realm of the Carl G. Jung (1875–1961), the Swiss-born psychologist, and his well-developed ideas surrounding the collective unconscious and synchronicities. As a researcher, it is safe to say that the patterns of information you interact with that are linked to your research are but a small sampling of the totality of information you engage consciously and subconsciously. All of your relations to finances, family, significant others, colleagues, community, state, national, cultural and social concepts, activities and events are additional patterns of information. Some of these are massive and are shared by literally millions of people today (and were embraced by countless of millions in prior years). In a recent review article, R Bobrow[2] in the Department of Family Medicine at SUNY- Stony Brook explored how emotionally connected individuals share a variety of traits such as obesity, happiness, and loneliness even if they have never met face-to-face. This is a type of social networking phenomena and provides evidence supporting the existence of a communal consciousness. Bobrow points out that this parallels what has been described among groups in the animal kingdom that effectively function as a single organism *(e.g., bees)*.[2] Marzouki *et al.*[3] go so far as to attribute part of social networking's collective consciousness capacity as playing a key role in a recent political revolution.

Synchronicity is defined in the Free Dictionary[4] as "the state or fact of being simultaneous or synchronous" or, alternatively, as a "coincidence of events that seem to be meaningfully related." The original treatise on synchronicity was developed by Jung during the 1950s[5] based on concepts he had introduced in lectures as early as the late 1920s.[6] Later, these were expanded on by Arthur Koestler in his book, *The Roots of Coincidence*.[7]

This chapter on synchronicity is not intended to detail the breadth or depth of the subject. There are numerous books on the topic, books dealing with Jung's evaluation of the topic, plus other interpretations of this subject. Instead, it is to make two points:

1) **Synchronistic events provide you with a personal measurement or indication that you are connected to a specific pattern of information.**

2) **Synchronicities may help you to identify, to understand better, and/or to have better command over patterns of information that you have engaged.**

In the event you are unimpressed by these two points consider the challenge of: (1) understanding the structure and boundaries of patterns of information you engage and (2) knowing precisely when you have connected to a specific pattern of information. These patterns can be massive in size and may interconnect seemingly unrelated pieces of your world. It is one of the reasons that a very common reaction to recognizing a synchronicity or coincidence is the feeling of awe or goose bumps.[8, 9] Synchronicities help provide a clear view of information terrains and tell you that you are connected NOW. Once you have that additional information, you are better able to perceive the pattern and to exert command over your interactions with the pattern. That is a rare opportunity and one not to be missed. Instead of being mystified by synchronicities and trying to decode what they might mean within the cosmos, we can take advantage of their occurrence, look for more of them, and use synchronicities to gain better command over our interactions with the relevant informational patterns.

Jung's own encounter with synchronicity occurred when he was sitting in a darkened room listening to a woman tell of her dream in which she was given a golden scarab. Behind him Jung heard a tapping on the window-pane. Upon opening the window he caught the "tapping" insect. It was the closest thing phylogenetically to the scarab beetle of the women's dream that lived in Jung's locale. It was also trying to enter a darkened room, which was completely uncharacteristic of its usual nocturnal habits. This happenstance made quite an impression on Jung. As emphasized by Mansfield,[10] Jung considered synchronicity to be nonlinear. He felt it could not be explained by traditional cause-effect relationships.

In his monograph, *Synchronicities an Acausal Connecting Principle* originally written during the 1950s, Jung[5] detailed the concepts and examples of acausal but meaningful coincidences. Yet, according to Joseph Cambray,[6] Jung's treatise almost did not happen. Jung had been so uncertain as to the way forward. Eventually, he was helped by his communication with Wolfgang Pauli, before the earliest comments on synchronicities were extended to a completed published theory. One tipping point in his moving forward with the synchronicity monograph appeared to revolve around his own additional experiences with synchronicity. One of the most profound experiences for Jung occurred as he was strolling by a lake where he was staying. Jung found a snake that unsuccessfully tried to swallow a fish. Both animals had died in the process frozen in their mutual encounter.[6] According to Cambray,[6] Jung interpreted this to mean that efforts to somehow meld Christianity (the fish) and alchemy (the snake) were producing a fatal outcome to both. It was such powerful imagery and related validation for his ongoing work that he followed the path toward producing the monograph on synchronicity.[6]

Not surprisingly, we have had our own encounters with synchronicities including a few connected to the preparation of this book and the related Cornell University course. One event occurred just as I was making preparations for the debut of the course and having some major self-doubts as to whether my college would see the course as a worthwhile effort in an university setting. But just then, I learned that a major college symposium on creativity (discussed in Chapter 5) was preceding my new course by a few weeks and would include my own college dean as a panel member. To phrase the issue another way, if I were concerned about how my college might view teaching enhanced creativity to students, what better answer could I get than to have my own dean involved in a program specifically designed to teach creativity to students? This was both synchronous and very useful for my own peace of mind. Playing off of this synchrony, the college then did a newspaper article and an alumni article on my new "creativity" course, which led to additional presentation opportunities. Finally, as the course was in full swing, I stumbled across a flyer for a major lecture from one of the coauthors, Rebecca S. Robbins, of Dr. James Maas' latest book, *Sleep for Success*.[11] While this is a wonderful topic, it is not one you would regularly expect to see on the special seminar lists of Veterinary Colleges. In fact, it is probably a first for my college. It was also timed to occur in synchrony as I would be covering similar material in the new course, and we would be revising our first-draft chapter on using sleep as a tool for creativity in this book. My synchronicity-associated awe was in full swing.

Clearly, the events were programmed for other quite useful purposes than inducing my awe. But these unexpected events were pivotal in signaling that the way ahead for teaching creativity at Cornell was paved with support and was a leap-of-faith well taken. The pattern was large and much clearer than only a few weeks before these events popped up.

JOURNAL IT

With synchronicities as with many pattern interactions, you are often dealing with large patterns where the nodal points in your perceptional awareness can appear to be non-linear or illogical. These coincidental events can be easily forgotten. As described with the prior personal example, it may take more events and more time before the pattern becomes more obvious to you. For this reason, it is very useful to record these synchronous events. Keeping a journal of these happenings is something that has been suggested by numerous authors who have written on the topic including: Deepak Chopra,[12] Frank Joseph and Dale Graf,[13] Paul Bishop,[14] Michael Scammel,[15] and Trish and Rob McGregor.[16] The journal becomes part of the pattern recognition process. Much like practicing noticing anomalies can lead to seeing more anomalies, entering synchronicities in a journal is likely to lead to your noticing more events for inclusion.

Deepak Chopra[17] recommends five major steps to increase the role of synchronicities and their utility in your life.

1. Record all coincidences in a journal as a way to avoid ignoring them. He even suggests placing them in different categories by the significance of the coincidence (e.g., tiny, small, medium, large). We suggest you create your own scale. Chopra believes that keeping a journal or diary of coincidences actually nurtures them.

2. Recapitulate the event. Chopra advocates replaying the event much like a movie. Setting this up just before going to sleep can be beneficial.

3. Be objective in viewing your coincidental movie. This is what we term the neutral observer effect as is discussed in our later chapter on this topic. You want to see the events and any connections from a nonpartisan perspective. It is a different and useful vantage point. If you remain immersed in the pattern, it can be more challenging to see the panorama of the pattern.

4. Update your journal entry of the event by adding the observed movie impressions.

5. Pose fundamental questions to yourself (e.g., What do I want to do with my life?) and watch to notice the interconnections of your recorded synchronicities and answers to those questions.

The examples of the questions Chopra[17] provides are all open-ended. This fits the model also suggested by Bartlett[18] in valuing the use of open-ended questions.

Synchronicities are a great way to access information surrounding your most fundamental questions. Note that when you set it up this way you are beginning to use one of our personal mantras: your game, your rules.

FEED THE JOURNAL EXERCISE

This exercise will not take much time, and the activities are distributed over the course of a week. But taking the initial steps to record your synchronicities can pay significant, long-term, personal dividends.

1. If you do not have one, start a journal of your noticed coincidences. Make your first entry from memory of some prior event. This has provided a symbolic first "meal" for what will serve you as a type of living companion: your journal that details synchronicities as an ongoing element in your life.
2. The next day, open the journal and read your entry. Enter another coincidence that you can recall. It does not have to be recent but may have occurred in your distant past. Close the journal. You have again fed this new companion.
3. Each succeeding day, read all the prior entries and make one new entry. If needed, wait a few minutes until something comes to mind to enter. With continued nourishment, the new companion can thrive.
4. After seven days read all the entries. You can choose to continue your journal if helpful. You have allowed this new companion to grow and be sustained for a week. Your recognition of coincidences is a new element of each day. This can continue and flourish as a pattern much as your journal is now flourishing.

THE NEED TO KNOW?

In a recent Psychology Today article, "Magical Thinking, Delusions, or Synchronicity? The Stories We Tell About Mysterious Phenomenon Shape Our Reality." Kim Schneiderman[19] raises an important point. She discusses whether knowing the basis of synchronicities is more important than their perceived usefulness. Schneiderman[19] uses the example of a client who perceived that the number 14 was somehow interconnected with a path toward more fulfilling relationships. The number 14 always showed up for him when he was taking a step that subsequently turned out to be useful. On the surface the numerical-relationship connection seems illogical. Any connection is certainly not linear. It was not particularly important to Schneiderman whether her client's interpretation of the source of this connection (via his grandmother who wished him well in relationships) was correct. She points out that some would label this magical thinking, others synchronicity and still others, a delusion. But in the end it is a story that becomes part of a person's reality. The significance was that this client's apparent synchronicity had been noticed, was part of his perceived reality, and might well have been working to his benefit.[19]

We do not attempt nor do we intend this to be a treatise on Jungian psychology and behavior. But for the researcher it can be useful to understand the broader landscape of informational patterns and to use this understanding to an advantage when it comes to broader opportunities for creativity and innovation. To this end we have a brief exercise that may help you to measure your boundaries in thinking about patterns of information and their engagement.

NEED TO KNOW EXERCISE

If you were given an object, at no cost and minimal risk, and were told: (1) it could aid your next project at work, (2) that it required no real action on your part, (3) that the processes used surrounding the object were uncertain, not-yet-established and/or even debatable..... would you still consider using the object?

This question sequence can be flipped if it is more useful: What is the minimum you would need to know about the object before deciding to use it? (1) Do you have to know exactly how it works? (2) Do you have to know that a current understanding of its mechanisms conforms with your scientifically-held paradigms before using it?

How you answered these questions may help you to determine your degree of openness for accessing and using pre-existing patterns of information.

SUMMARY

Noticing synchronicities in your life provides a calibration opportunity for your connections to patterns of information you have previously observed or have shared with others. Based on the experience of several authors, recording your synchronous experiences appears to increase the frequency of noticing these occurrences in the future. As noted by Schneiderman, you may choose to label these experiences as magical thinking, synchro-nicities, or moments of delusion. Regardless, they can be highly useful and represent another tool for enhanced creativity and innovation. We encourage you to embark on the process of noticing more synchronicities in your future as a scientist.

Chapter 14

Sandman Science: Science Sifting During Sleep

We are such stuff as dreams are made on... — William Shakespeare[1]

One of the prime tools for solving scientific and general roadblocks is sleep. Most likely you recall some time in your life when you went to bed facing an immovable roadblock that had defied resolution only to awaken in the morning with a surprisingly workable solution. Many times your morning solution was not even on the table the previous evening. In fact, Mason Cooley, the noted American aphorist and university professor suggested that "When you can't figure out what to do, it's time for a nap."[2] We absolutely agree. The age-old suggestion "why don't you sleep on it" has an actual scientific basis to promote a different kind of problem solving.

How does sleeping on it or a fuzzy-mind state work in solving problems? There are ideas concerning what appears to happen. First, there appears to be a selection process among factoids or concepts that were sequestered in consciously-obscure thought. These are thoughts that had been buried in a folder labeled as not that relevant to the issue at hand.

I (RRD) have a first-hand example to describe involving this chapter. One evening well into the drafting of this book, I attended a faculty-student reception. It provided an opportunity to discuss this book-in-preparation and the related lecture materials with a colleague whom I almost never see and had never talked with on anything at any great length (one of the benefits of such casual interactions). When I walked up, he was discussing a seemingly intense and challenging hobby, which was physically-grueling bicycle rides. This piqued my interest and during the course of the discussion I asked him if he found that he got scientific insights into his own research during his prolonged bicycle rides. He responded yes, but.....it was sleep that was his best tool for breakthroughs. Furthermore, it seemed to be a family tradition as sleep had been a type of creativity salve for his father as well. After a day spent in less-than-productive, left brain-focused head butting, my colleague would awake with precisely the piece of information he needed to overcome roadblocks.

You might think that this conversation led immediately to my taking up the gauntlet to write a sleep chapter. Well not exactly. I had filed away my colleague's story as interesting and not the first time that I had heard about or experienced sleep having such a side benefit. But it was also a deviation from my then mission to obtain information on hobbies such as bicycle riding. That was where my linear focus was directed. So I did my best to at least temporarily overlook the testimony to the creative benefits of sleep. But then came my own night's sleep and I awakened at an all too early hour incapable of doing anything else but beginning to write this chapter on.... sleep (while counting the drips of the awakening coffeemaker). In fact, there was little else I could do that day except follow the impulse to write about sleep as I tried to move beyond my sleep-deprived state.

It was as if some multi-dimensional prospector, like the one on the cover of *Science Sifting,* had used my sleep time to sift through all my informational and experiential folders as well as information outside of me to find the most useful combined pieces of information — the golden nuggets concerning sleep. The science sifting was done and I was ready to cash in those nuggets. This was not the first time nor the only way an unplanned book chapter got inserted into our shape-shifting table of contents. What is interesting about the "awaking knowing" that comes from sleep is that it is not simply a second opportunity to view a buried factoid. It also has an element of increased discernment. This information is often delivered with a prioritization or relative ranking above and beyond all other factoids or concepts that you can consider. And even better, it is sometimes quite difficult to explain logically why the post-sleep "awaking knowing" should have a top priority. Sleep-associated information is simply moved into our foreground and everything else becomes out of focus and blurred in perception by comparison.

The use of what we call science sifting during our sleep is probably one of the more widely recognized tools for problem solving in research and technology as well as in virtually every other aspect of our lives. It has been used by researchers, artists, world leaders, the penultimate innovators and most famous inventors across several centuries. What we will consider in this chapter is not only that sleep is a useful tool for creativity but also that this tool has the potential to be manipulated and directed toward specific problem-solving with proper preparation.

SLEEP FOR ENHANCED INNOVATION

One of the things that sleep allows you to do is to access conceptual imagery. Sometimes vivid images appear to us during sleep, and these are part of awakening recall. The famous American painter Andrew Wyeth once said "I dream a lot. I do more painting when I'm not painting. It's in the subconscious" (quoted in Meryman[3]).

Psychologist Marieke Wieth of Albion College did a study[4] to determine the peak performance for individuals to successfully solve analytical vs. insight-based problems. She found that fuzzy awareness was useful. Her results suggested no specific time relationship for solving basic analytical problems.[4] But that was not the case when creative problem solving was needed. If creative insights were required, the best windows for obtaining solutions were during an individual's non-optimal times of the day. In other words, it was during the times we are least alert. Her interpretation was that during our peak alertness times, we use attentional inhibition to filter out any information we think will not be useful for our "intellectual" list of to-do tasks. Highly-focused "intellectual" attention on our tasks can keep us locked into a relatively uncreative mode for most of our waking hours (note this is often when we are at work).[4] It is only when our filters are down, as in a period of sleepiness, that we enter the window of greatest potential for creative breakthroughs.[4,5]

Studies from Germany examined the insightful prowess of two groups using a series of mathematical number sequence exercises.[6,7] One of the groups slept after completing a training trial while the other group did not. Both groups then returned to complete the remaining tests involving a series of number sequences. The group that slept finished in a shorter time. But what was most significant is that a higher percentage of sleepers found a new problem solving short-cut. They figured out a pattern that enabled them to complete the remaining trials faster. Importantly, when they awoke, they were not consciously aware of the pattern that gave them a short-cut. Nevertheless, they were able to access a shortcut that was subconsciously discovered during sleep. After the sleep vs. no sleep break, 59% of the sleeping group began using the short-cut whereas only 25% of the non-sleeping group found the short-cut. This suggests that sleep can represent a tool for accessing patterns of information that are not readily available via conscious, linear thinking.

WE ALL SLEEP

The linkage between sleep and creative problem solving has several implications. While most of us will sleep or at least try to sleep once a day, it means that achieving a less aware or a less highly-focused, linear-thinking state may be the real goal for enhanced creativity. In fact, most of the tools we describe in this book have at their core, the ability to bypass the intellect-driven, left brain-directed filters that lock us into limited analytical tasks. The fuzzy awareness state is our goal. Sleep is an important tool in the arsenal for opening the research creativity window. Since we all do it, we might as well let the multi-dimensional prospector start working on our behalf.

Among the tools we will discuss, you are likely to have the greatest familiarity with sleep. Beyond the fact you have done it regularly, most people will have had at least one prior experience in their life when they awoke with a bright, new idea. This is a useful frame of reference for what can happen when you, the researcher, relinquish control of the analytical tool you have worked so hard to bring to the workplace every day in every way. Just allow for these insights to show up.

But what if you have a deadline for the bright idea and need it now. Sleep works to provide solutions, but the results can be unpredictable and haphazard for most people. Can sleep-associated creativity be mobilized as needed? Can you have sleep-associated creativity-on-demand as a purposeful tool for greater innovation and creativity? Here we will provide some suggestions. However, specific approaches for using sleep to make a "creative space" are likely to require some individualistic tailoring. No single prescription should be expected to work uniformly for everyone. Within the information we provide, we suggest you find the menu that produces the best results for you. To examine this, we will use the similarities for effective science sifting that occurs between the tools of sleep and meditation.

SLEEP, MEDITATION, CREATIVITY AND OVERALL HEALTH

For increasing your vantage points and accessing patterns of information not available to your linearly-focused left brain, sleep has similarities to another tool we cover in the next chapter: the 30-sec meditation. It is a time when you can set aside the intense, sharp, linear focus that is a part of many researchers' 24-hour cycle and allow both information and new connections to come into your awareness.

Here we state we are not medical doctors and the following comments are not intended to be nor are they medical advice. You should always work with and consult with your physician in making health-related decisions. Sleep patterns can vary among individuals. While a minimum amount of sleep is useful for effective cognition and good health including effective immune function, precisely what is optimal for each individual (5, 6, 8, 10, 12 hours of sleep; continuous, broken into smaller units) and how that sleep fits into a personal cycle can vary particularly during different stages of a lifetime. Enhanced creativity would be a poor substitute for a sleep pattern that contributes to poor overall health. But this is unlikely to be a problem. It is far more likely that sleep patterns that enhance your creativity and insight are closely aligned with those that promote your good health. If your current sleep pattern provides you with few insights for overcoming road-blocks in your life, seeking a change could be doubly useful. You may be able to have a win-win in your life. You are likely to find that your problem-solving insights improve along with your health and vitality.

BEYOND A GOOD NIGHT'S SLEEP

Naps are not just for kids. They are a great idea for everyone. The scientific basis of the benefit of taking naps during all stages of life is discussed in some detail in *Why We Nap*, by Claudio Stampi.[8] A myriad of benefits are discussed including the observation that naps can be important in sparing the adverse effects of monophasic (e.g., a continuous night-time) sleep deprivation.[8] Practical examples of the benefits of naps are provided by William Anthony[9] in *The Art of Napping* where he concludes that naps are good for you and advocates napping whenever you have the chance including while you work. What we will emphasize here is that naps are not just a physiological refresher. Napping is a highly useful tool for opening your "creative spaces." They can: (1) help you bypass your consciously-established scientific roadblocks, (2) multiply your opportunities to enter "creative spaces" on a daily basis, (3) help you engage selective problem solving when linked with your frustrating work issues, and (4) feature the different depths of sleep states that allow you to interact with subconsciously-accessible information. Studies conducted by NASA have shown that naps enhance performance,[10] and this has helped to pave the way for broader acceptance of using naps in the workplace setting.

James Maas is a sociopsychologist at Cornell University and an internationally-renowned expert on sleep. According to Dr. Maas, a 20 minute nap in the middle of the day is worth more than the same time spent repeatedly hitting the snooze button to sleep a little longer in the morning.[11,12] He even advocates using the mid-day nap time to simply rest the eyes and brain and stresses those benefits as being well worth the time even if you cannot fall into sleep.[11] The concept and phrase "Power Nap" was first coined by Dr. Maas.

THE POWER NAPPERS

One of the significant observations about napping is that it is not just about accessing increased information. Equally important is the effect of napping on prioritized information or discernment. In their book *The Ten Faces of Innovation: IDEO's Strategies for Defeating the Devil's Advocate and Driving Creativity Throughout Your Organization*, Thomas Kelley and Jonathan Littman[13] discuss the importance of sleep for success including the role of the mid-day nap. They posit that the mid-day nap not only refreshes but has a curious ability to reduce information overload. Sometimes less can be more useful. This is critical for effective decision making as Malcolm Gladwell has pointed out in his book, *Blink*.[14] More and more information is not necessarily a helpful component for creativity if you are stuck in a rut regarding discernment. We previously mentioned

Gladwell's story of how an experienced field officer (Paul Van Riper) was able to out-think and out-maneuver the combined computational efforts of the US Joint Forces Command (JFCOM) in field trials called the Millennium Challenge. By using less information, having the ability to home in on the core factoids (here we would say the prospector's nuggets), and operating largely on instinct, Van Riper out-maneuvered the information-heavy (JFCOM). Not all information is equally useful and naps apparently help to separate the wheat from the chaff.[14]

A number of famous individuals in history are known to have used the nap for inspired thought even before Professor Maas gave it a name. Power nappers are said to include Thomas Edison, Nikola Tesla, Margaret Thatcher, John F. Kennedy, Winston Churchill, and Buckminster Fuller.

THE SCIENTISTS-INVENTORS

Albert Einstein

Einstein kept to a rather comfortable daily routine as has been described by his grandson, Bernard. After breakfast he loved to play the violin (while his housekeeper and secretary, Helen Dukas, did the dishes). Then followed work on correspondence with Helen. Later in the morning he was at the Princeton Institute for Advanced Studies. After lunch, there was a nap. Then it was back to the Institute. A light dinner would follow and then there was often more music before bedtime.[15] The importance of this to Einstein is also discussed by Ebert and Ebert[16] in their book on the creative mind in science.

Nikola Tesla

Nikola Tesla was reported to sleep only a few hours each night. But it also seems clear that he would often fall into daytime sleep. In his book, *Prodigal Genius: The Life of Nikola Tesla*, John O'Neill[17] describes how hotel employees would often find Tesla apparently transfixed in his room and completely oblivious to their presence the whole time required for cleaning the room. It seemed clear that if he slept only briefly during the night, he was in a trance or asleep probably more than once during most days. Lab staff also reported finding him napping more than once during a normal day.[18]

Thomas Edison

Like Tesla, Edison used to take great pride in, if not brag about, his ability to work long hours with little to no sleep. But also like Tesla, it was clear that when Edison meant little sleep he was only accurately referring to standard nighttime sleep. Edison was not without sleep if you included his frequent daytime naps. Martin describes how Edison kept a folding bed near his lab and used tactical naps as a means of "creative hypnagonia."[19] One famous example of Edison's napping occurred in 1894.[20] Edison was scheduled to welcome a celebrity to Menlo Park. A European strongman touted as a modern-day Hercules, Eugene Sandow, had travelled to be filmed using Edison's Kineoscope. But when the dignitaries arrived, one of Edison's assistants had to greet them, apologize and proceed with the filming in Edison's absence. As lab assistant William Kennedy Larie Dickson explained, Mr. Edison was taking a nap. Sandow did get to meet Edison at lunch later that day.[20]

Leonardo da Vinci

In his book, *Sleepfaring: A Journey Through the Science of Sleep*, James A. Horne[21] describes how Leonardo da Vinci used a cycle of short naps (estimated at approximately 15–20 minutes in length) at regular intervals particularly when he was working on the large murals with paint drying time constraints. In fact the urban dictionary (http://www.urbandictionary.com/) refers to a series of short intermittent naps as "Da Vinci sleep."

POLITICAL LEADERS

Winston Churchill

Winston Churchill was famous for dividing his day into a noontime session of either painting or cards and a two hour afternoon nap. He even kept a bed in the House of Parliament specifically for this purpose. According to Leonard Spencer-Lewis,[22] Churchill's naps were sacrosanct, and he gave instructions that they were only to be interrupted if German forces landed in England. In fact the author goes on to refer to naps as taking a "Churchill." In his book, *The Quality of Leadership*, Michael Hansbury[23] quotes Churchill as saying that "You must sleep sometime between lunch and dinner, and no halfway measure. Take off your clothes and get into bed. That's what I always do. Don't think you will be doing less work because you sleep during the day. That's a foolish notion held by people who have no imagination. You will accomplish more...."[23] Hansbury even divides naps into nano, micro, mini and macro categories.[23] When you read about the 30-sec heart meditation exercise later in this book, that would fall into the same time allotment as Hansbury's nano-nap (a 30-second to 2 minute nap).

John F. Kennedy

John F Kennedy used a post-lunch power nap as a means to being more productive during the lengthy day. Jacque Onasis described how Kennedy regularly used to change into his pajamas for a 45 minute afternoon nap.[24] Kennedy had been impressed with Churchill's effective use of afternoon naps and had begun doing the same while still in the Senate. He continued this schedule during his White House years.[24]

Lyndon B. Johnson

This presidential nap trend continued with Lyndon Johnson who used what Parker describes as the "Johnson Double Shift."[25] It was an eight-hour shift followed by a nap and change of clothes and a 5–6 hour second shift. Like Churchill and Kennedy before him, Johnson would change into panamas with a nap of 45 minutes to an hour. [26] In his book on the Kennedy tapes, Michael Beschloss[27] quotes Jacqueline Kennedy as having persuaded Johnson to start taking afternoon naps. She stressed how it had "changed Jack's whole life."[27]

Ronald Reagan

The trend did not stop there with American political leaders. Ronald Reagan also used afternoon naps on a regular basis. In fact, in his edited book on the Reagan diaries, Douglas Brinkley[28] described several entries concerning sleep and naps including an incident where Reagan, upon awakening, drafted a letter to Brezhnev and was particularly pleased to have captured those thoughts regardless of whether the letter was ever sent.[28] This is why, in our opinion, the impressions you have upon awakening really are among the best synthesis of information that you can tap into.

William J. Clinton

William Clinton mastered the use of naps beginning in the 1974 political campaign and continued this practice through his White House years. Ann McCoy an administrative assistant, asked him one day how he did it so effectively and the reply was "I just imagine a big hole in the back of my head, and I focus on that."[29]

THE ARTISTS, COMPOSERS, AND MUSICIANS

Highly creative people use naps as a way to either enter into or to sustain their maximally creative periods. Restak[30] advocated the use of short naps as a restorative solution to maintain high levels of concentration during creative efforts. He described the behaviors of professional musicians and their effective practice time. Naps could be used to extend the daily duration of effective practice.[30] This is precisely what has been described in the routine of Malcolm Lowe, violinist and concertmaster for the Boston Symphony Orchestra.[31] In a somewhat dramatic display of nerve, some orchestral musicians go to the extent of taking short naps during long periods of rest in concerts.[32] We suppose if you were expert at precision naps, a performance of Wagner's later operas (e.g., Die Gotterdammerung, Parsifal or Die Meistersingers) could become as refreshing as a weekend retreat.

Salvador Dali

Noted Catalan artist Salvador Dali not only took micro-naps but was quoted as believing that it was one of the pivotal points to his becoming a great artist. He spoke of "slumber with a key"[33] believing that his micro-naps revived both his physical and his psychic being. Using this technique to grab the benefits of abrupt sleep, Dali would sit with keys held loosely in his hand over a plate. Upon falling asleep the keys would drop and awaken him. He found he got the creative nap-associated benefits he sought without excessive time investment.[34]

Wolfgang Mozart

While Mozart is most noted for his feverish composition of music, his capacity to nap without losing a beat was also part of the creative equation. On the eve of the premier of his opera *Don Giovanni* he was still writing the overture. His wife kept him awake as needed with chatter, then he would nap and the sequential process was repeated as the completed pages were run by messenger to the orchestra in rehearsal.[35] Note we are not necessarily recommending this specific approach as being health-sustaining, although one could envision it might occur at some point during a researcher's career if one substitutes a grant proposal for an opera.

Giaccomo Rossini

Rossini was noted for not only being one of the most prolific composers ever during his period of work, but also for having the distinction of being one of few who knew when and how to retire so as to outlive his contemporaries. A description of his schedule during his operatic years in Paris stated that Rossini would eat an early dinner, nap, socialize and then return to bed.[36]

Igor Stravinsky

In E.W. White's book, *Stravinsky: the Composer and His Works*,[37] a holiday visitor of 1947 described Stravinsky's study in Los Angeles during the height of the composer's productivity. The carefully planned room had two pianos, two cupboards, two tables, two desks, and a strategically-placed couch that Stravinsky used for his afternoon naps. The description went on to include details of Stravinsky's facial expression when sleeping and his gentle snoring.[37]

Guiseppi Verdi

Verdi owned a countryside estate at Saint' Agata from 1851, the year he composed Rigoletto, until his death in 1901. It was here that he lived with his wife, Giuseppina Strepponi, and where he composed the majority of his greatest works. As part of his daily routine, Verdi usually rose early but by afternoon took a 1–2 hour nap.[38]

HYPNAGOGIC NAP

In their book, *Take a Nap, Change Your Life,* Mednick and Erhman[39] refer to the pre-deep sleep stage as the hypnagogic nap, or skimming along the surface of sleep. It is the just pre-sleep or light-sleep state where imagery often arises. It is also likely the same experience that Salvador Dali enjoyed during his micro-naps. Even a brief diversion into the sleep state allows your conscious mind to step aside. You are then free to sift through information that may be completely unavailable to your conscious mind. Stickgold *et al.*[40] showed this effect in a study of subjects with anterograde amnesia. The researcher found that these subjects could not form a declarative memory of the game Tetris even if they had been playing it and only momentarily glanced away from the screen. However, if they were allowed to enter a light sleep and were then awakened, they reported dreaming of falling Tetris pieces.[40] Clearly, for these subjects the information was available to them but only as an image capture. What was needed was the prospector's sifting and an awareness opportunity.

OPTIMIZING THE USE OF SLEEP
FOR SPECIFIC TASKS

One of the big questions surrounding sleep and creative insights is whether we must passively wait for this to show up or can activity increase the likelihood of useful creative work during sleep? Can task-oriented work be facilitated beyond simply thinking of a problem before a good night's sleep or a power nap? Some evidence suggests that prompts can be given during one's sleep that makes the occurrence of specific, creative, task-oriented problem solving more likely. For example, Ritter *et al.*[41] used what was termed task reactivation to enhance the incidence of task-directed creativity during sleep. The researchers studied groups of individuals who were conditioned with an odor introduced with a problem prior to sleep and then exposed to no odor, a different odor, or the same odor during sleep. When the group that slept with the conditioning odor awoke and were assessed, they were found to be more creative and also better able to select their creative ideas than the other groups.[41] This suggests that we can actively plan for creative problem solving during sleep.

THE ODOR PRIMED SLEEP EXERCISE

1. Identify a problem or issue you would like to work on during your sleep.
2. Anytime up to two hours before your bedtime, sit with a problem or issue written out in front of you and think about it while you have a familiar odor nearby (e.g. vanilla, citrus, air freshener). 5–10 minutes is long enough.
3. When you are ready to sleep at night, place the same odor source within nine feet of your pillow (on your bed stand.)
4. The next morning look at the problem and notice if by noon you have any new thoughts concerning the problem/issue that are useful. Has there been any progress to overcoming a roadblock or gaining new insights?
5. We suggest you enter what you notice in a journal (even if you noticed nothing).

You can try this more than once during the week if it does not disturb your sleep.

SLEEP DISORDERS

In any chapter on sleep and its wide range of benefits, the topic of sleep disorders is also worth a few lines. Sleep problems are among the cadre of chronic conditions that have been increasing in prevalence in recent years. The US Centers for Disease Control and Prevention recently estimated that more than a quarter of the US population reports at least occasionally not getting enough sleep.[42] Dietert *et al.*[43] found that sleep disorders are connected to several patterns of chronic diseases and conditions that are driven by improper regulation of inflammation. In the circularly-related connection, persistent lack of sleep can reduce host defenses and increase the risk of certain diseases. And increased inflammation from certain chronic diseases can contribute to loss of sleep. Quality sleep is clearly an important consideration for health and well-being for everyone including the research scientist. One of the keys to a healthy, productive and maximally creative research career is to establish a sleep pattern than is optimized for your body.

SUMMARY

We all sleep and it is perhaps the most underutilized tool for creativity that we have readily available. Sleep provides an opportunity for problem solving without the linear thinking and filters that can inhibit our most innovative possibilities. Beyond the access to information, sleep and power naps carry with them a significant capacity for discernment. We awake not simply with useful information but also with effective prioritization of available information. A myriad of the most innovative scientists, artists, and world leaders have used naps as a tool for innovative breakthroughs. Even a brief 20 minute nap can both refresh our bodies and open our awareness to new solutions. Reactivation techniques suggest that we can sleep with intent when it comes to problem solving. That makes the title of Cornell Professor James Maas' recent book, *Sleep for Success,*[12] all the more appropriate for our goal of enhancing creativity in research.

Chapter 15

Meditation: Ohming Up or Getting Down with the Real You?

Tell your son to stop trying to fill your head with science — for to fill your heart with love is enough. — Richard Feynman[1]

Among the chapters of *Science Sifting* you are presented with tools designed to provide you with more vantage points. Some of the tools are messy as when you did the airplane exercise or littered your office with balled up pages of your grant proposal. Others could be mildly disturbing as when you deliberately read dramatic news stories to calibrate the effects of drama on your body. Some require physical actions as with the Walk-Around exercise. Still others can be noisy like with Dali-sleep where you might deliberately make a loud noise to startle yourself awake. These are fine if you don't mind a little exercise and possibly breaking a lot of dishware. They work. But the good news is that our arsenal of tools also contains reduced stress and less noisy exercises as well. A major approach for enhanced creativity is meditation.

Before you start running for the exits figuring that this book purportedly written for scientists has entered some New Age wonderland, it is worthwhile to mention that meditation is a major programming component of The National Center for Complementary and Alternative Medicine (NCAM) at the US National Institutes of Health (http://nccam.nih.gov/health/meditation/overview.htm). They fund research projects into the health benefits of meditation, and these funded programs have already produced some impressive results as will be discussed later. Additionally, if you look around the workplace, you have a better than 50% chance that your institution has an employee program that includes meditation as part of its healthy-workplace initiatives. There is a reason for this. It works. Further support for the growing impact of meditation is evidenced in publications such as *The Complete Idiot's Guide to Meditation*,[2] which was released in a second edition over a decade ago.

Between us, we have led group meditation exercises during research presentations at a US Federal Agency (the USDA), various Cornell University department research seminars, the Cornell BioMS 4400 course, and off-campus workshops. I (RRD) also use many of the tools from this and other chapters in the book to enhance additional conference and classroom experiences such as those connected to our Doctor of Veterinary Medicine curriculum. It is clear that the learning processes and even the nature of the questions that arise in meetings, conferences and classrooms are strikingly different when these tools are used. In this regard, meditation is both personal and interpersonal. What we will argue in this chapter is that the same procedures designed to reduce stress, enhance efficiency, and provide health benefits for employees can be applied directly to increase the creative spark of the research effort itself.

This chapter provides another tool for enhancing creativity in your research career. But we will stress that meditation might be called a super tool. Part of the reason is that it serves as a tool by itself; but it can also help to make the other tools work more effectively and consistently to your benefit.

Meditation is not a single procedure or approach. In fact, there are many different forms of meditation, and there is some evidence that not all forms of meditation produce identical physiological effects. Forms of meditation can differ as widely as different areas of science, and we are unlikely to think of molecular biologists as fully-interchangeable with field ecologists. The same cautionary note should apply to different traditions of meditation. We will not come close to presenting the range of traditions of meditation within this chapter. Instead, we will focus on three types of meditation where we have more direct knowledge and/or direct experience and that are quite accessible: mindfulness meditation, heart-based meditation, and transcendental meditation.

The English word meditation is derived from the Latin "meditatio" from the verb "meditari" meaning "to think, ponder, devise, contemplate." For our purposes, we will emphasize meditation as a tool for devising a new way to ponder. While many authors discuss meditation as being all about the mind, it is important to realize that usually refers to your body's knowing mind. This is definitely not the same as your brain. We will emphasize that meditation is less about the brain's contribution to mind and more about your holistic body-mind relationships and consciousness layers of information. In fact, for scientists who have spent much of their day in a focus, focus, focus mode of operation, meditation provides a useful route for getting out of the brain and into the body's broader information terrains.

Meditation does not have to be an elaborate event as you will see. In fact, Puddicombe[3] stresses that one can learn to meditate just about anywhere, and it does not have to take long to be effective. It is not about being goal-oriented. Instead, it is about being in yourself in the moment. As Puddicombe states "when it comes to meditation.... the goal and the journey are the same thing."[3] Meditation is simply the art of getting out of one's way and out of one's well-conditioned brain.

MEDITATIVE TRADITIONS

Historically, the practice of meditation has been connected to many different religions including but not restricted to Buddhism, Christianity, Hinduism, Islam, Judaism, Sikhism, and Taoism. If meditation has a specific spiritual or religious connotation for you, we encourage that continuation. If you view meditation as something apart from a specific religion or faith or as secular, there is a tradition for that as well. We encourage you to investigate meditation as a useful personal tool.

In discussing meditation as a tool, we will deviate from many presentations on the subject. We are far less concerned with the how and why of meditative practices than we are with the increased possibilities that result from meditation. For this reason, we will not spend much time devoted toward efforts attempting to explain theories and experimental results concerning precisely what happens during meditation. It is a subject of interest to many and is significant. However, that is outside the main purpose of this book.

What does meditation have to do with research? Being able to willingly shift in and out of different operating modes can be very useful for the researcher. In 2007, a national US survey[4] reported that 9.4% of adults had used meditation as mind-body medicine in the past 12 months compared with 7.6% in a survey conducted in 2002. So the practice of using meditation is reasonably well–represented today even in the United States. Your applying this in your research efforts will simply add to the growing number of people who look for non-pharmacological routes for increased awareness.

MEDITATION AND YOUR PERCEPTIONAL AWARENESS

It is likely that if you currently have a research challenge or block, you did not get there through the path of meditation. As a result, use of meditation is unlikely to be a repeat of the same running headlong into a brick wall process that resulted in a roadblock. But you might like additional assurance that meditation could be more useful than simply telling you that it is something different.

There is evidence suggesting that cognitive flexibility is a hallmark of meditation. For example, Greenberg *et al.*[5] reported that mindfulness meditation practices reduced cognitive rigidity based on the performance of subjects in a problem solving test. The researchers suggested that meditation gave participants a "beginner's mind" when it came to the task and helped them to avoid being "blinded" by past experiences.[5] Obviously, this is precisely the aim for using meditation as a research tool. It enables the researcher to approach a problem with a completely fresh, first-time look. Additional studies also provide support for the tenet that mindfulness meditative training aids both attention as well as cognitive flexibility as it pertains to problem solving.[6–8] Meditation has also been reported to increase the structural plasticity of white matter based on *in vivo* MRI assessment[9] and appears to improve the self-regulation of attention.[10] Finally, Ostafin and Kassman[11] also reported evidence for a connection between mindfulness meditation and creative problem solving.

There are as many prescriptions for how to pursue effective meditation as there are cultures and traditions that use it. Many meditative traditions have strict protocols including step-by-step procedures that must be followed, details of required posture, vocalization, and duration. From our vantage point, we respect the diverse spectrum of meditative protocols but adhere to none as preeminent over all others. Our primary interest is to help you consider meditation as one of your tools for increasing your vantage points. If you develop a meditative procedure that involves spinning like a whirling dervish while singing a Beatle's song and performing walk-the-dog with a yo-yo even as you twirl, it is fine with us (as long as you do not expect us to do it with you). If you have a meditative protocol that you can do regularly, it does not produce bodily harm to you or others, and it works for you, three cheers. If you don't already have a personal process in place there are numerous resources available. Among these is the University of California at Los Angeles (UCLA) Mindful Awareness Research Center (MARC) based in the Semel Institute at UCLA. They have a number of meditative exercises and programs available via the internet at: http://marc. ucla.edu/. Other prominent university-based programs also exist. Among these are the Center for Mindfulness in Medicine, Health Care, and Society at the University of Massachusetts Medical School, The Penn Program of Mindfulness at the University of Pennsylvania and The Center for Mindfulness at the University of California at San Diego. In addition to these resources, we will describe some features and short-cuts of meditation that may be useful for gaining new perspectives.

One of the critical keys to meditation is the capacity to observe real-time from a nonjudgmental perspective. Here is an example. Mindfulness meditation training has been shown to have positive effects on reducing pain. But what does it actually alter? Recently two University of Manchester researchers, Brown and Jones,[12] reported evidence that meditation reduced the "anticipatory and negative appraisal of pain."[12] In other words, it may not have reduced the factors that led up to the pain, but altered the perception of those factors, instead. We would argue it took the individual from a well-learned linear relationship between signals, fear, and discomfort to a nonlinear association where the signals did not have to produce fear or discomfort. It is where an engrained flight-or fight type of response was no longer the default operating mode.

ADJUNCT BENEFITS

Before discussing three specific types of meditation, we point out that increased perceptional awareness is not the only benefit attributed to meditation. If you incorporate meditation as a new tool for research innovation and creativity, you are likely to see specific physiologically-based benefits. A recent study among a highly-stressed population, caregivers of dementia patients, found that brief daily meditation, in this case yogic meditation, was associated with an increase in immunoglobulin gene transcription and reduced proinflammatory cytokine production.[13] Considering that misregulated inflammation is at the root of many chronic diseases, this is a useful outcome.[14]

Meditation is thought to alter your relationship to pain. Supporting results were reported from the Wake Forrest University School of Medicine.[15] These investigators found that prior meditation training reduced the level of unpleasantness of pain by reframing how you perceive pain events. Researchers in Oslo, Norway[16] examined the effect of 20 minutes of nondirected meditation on several physiological parameters including heart rate variability (HRV). They found that among middle-aged men and women, meditation increased the variability of heart rate in two different bands of frequencies.[16] This is important because you are at a reduced risk of several diseases when your heart rate shows more variability.[17, 18] The investigators concluded that meditation may contribute to a reduced risk of cardiovascular problems.[16]

THREE EXAMPLES OF MEDITATION

1. *Mindfulness Meditation*

Mindfulness meditation is a calm-based, increased awareness process designed to bring one closer to present-moment reality. Karen Kissel Wegela[19] describes mindfulness meditation as teaching us how to be unconditionally present. It is not so much about shifting you to something different as it is noticing where you are in the moment. This meditative practice focuses, in turn on body, breath and thoughts. Body is seen as being in relationship to environment, breath is viewed as it exists in the moment, and thoughts are noticed as something that has occurred but without any judgment. Wegela[19] emphasizes that mindfulness meditation contains an aspect of noticing you as you are but from the point of view of a neutral observer. Note that we have a chapter on becoming the neutral observer later in this book.

One of the major proponents of mindfulness meditation is Jon Kabat-Zinn, Professor Emeritus and founding director of the Center for Mindfulness in Medicine, Health Care, and Society at the University of Massachusetts Medical School. Dr. Kabat-Zinn is credited with taking meditation into medicine in a way that has enabled it to grow and flourish as a health-promoting practice. His teaching of mindfulness meditation for stress and pain reduction began as early as 1979. In his book, *Mindfulness for Beginners: Reclaiming the Present Moment — and Your Life*,[20] Dr. Kabat-Zinn defines mindfulness as "awareness, cultivated by paying attention in a sustained and particular way on purpose, in the present moment, and non-judgmental."[20] He further describes it as "much ado about what might seem like almost nothing but which turns out to be just about everything.....it contains a whole universe of life-enhancing possibilities."[20]

In speaking of the great teachers, Kabat-Zinn stresses the importance of never losing the "beginner's mind."[20] That is where the creativity resides. Each moment is new and so too can be the beginner's mind in that moment. Throughout this book, we also stress the importance of returning to your childhood roots of openness, curiosity, inquisitiveness, and non-judging observation.

2. Heartspace-Based Meditation

Among the forms of meditation that are used is a cadre of exercises aimed at getting you out of your head and into your heart. The heart area is a potentially useful location to place your awareness during these meditations. Despite the recent popularity of heart-based meditation, it has its roots in antiquity. Some early mentions of heart-based meditation its roots are with both Sufi and Christian "prayer of the heart" traditions.

You can find a plethora of websites, books, and CDs surrounding heart-based meditations including exercises and a range of potential applications.[21-25] Among the devotees of heart-based meditation are Howard Thurman, Gregg Braden, Richard Bartlett, Irmansyah Effendi, Tara Brach, and Harshada Wagner.

There are several reasons why a focus on the heart as a meditation destination can be a useful choice for the scientist. Getting out of one's head and in better connection with the whole body is a novel diversion when we spend so much time engaged in left-brain focus. Compared with other parts of the body, the heart produces both a torsion field and a significant electromagnetic field.[26] Finally, the heart-brain interconnection is very strong and views of what constitutes the anatomical center of the mind via this interconnection are likely to differ depending upon one's medical specialization.[27] For these reasons, directing your attention to your heart area could be a very useful experience. Efforts to examine the impact of heart-based meditative protocols in childhood learning have been underway.[28]

In the next section we will provide you with a personalized heart-based meditation exercise. In the literature, the nearest descriptions of what we do are probably that described by: 1) Bartlett[24] when he discusses dropping into the field of your heart, 2) Seaward[29] in the Taoist-tradition of heart meditation and 3) Vas[30] who describes Khan's heart rhythm, meditative exercise.

The following heart-based exercise is one we have taught in Cornell University seminars and classes, on-site at a US Federal Agency, in community groups, and via a national webinar. In final preparation, we recognize that some forms of meditation are conducted with intense ritual, and the sought-for results can take days, weeks, years or even decades. We don't have that kind of time and suspect you may not have meditation week entered into your day-planner either. So our version of meditation takes approximately 30-seconds. Remember, in the world of nonlinearity, the amount of time invested is unrelated to the results seen.

30-SECOND HEARTSPACE MEDITATION EXERCISE

Here we go.

1. First look around and notice what the room you are in is like, and how your body feels. This is your pre-meditation calibration.
2. Sit or stand whichever you desire. Either works. You can close your eyes if you like.
3. Relax your body and arms, rolling your shoulders back and down slightly.
4. Let your chest and stomach protrude slightly as it is your chest-midsection that will occupy your awareness.
5. Take your awareness and move it from in front of your eyeballs to inside your head behind your eyeballs.

(Continued)

30-SECOND HEARTSPACE MEDITATION
EXERCISE (*Continued*)

6. Let your awareness step onto an elevator and go down into your body, stepping off at your heart level.
7. Now it is like you have a spotlight shining out of your chest. For the next few seconds, notice what the room and the air are like. What does it feel like inside your body? This is your in-meditation calibration.
8. Now reverse the whole process riding the elevator back up and taking your awareness from behind your eyeballs to in front of them.
9. Now look around the room and notice how it feels to you. Is anything different? This is your post meditation calibration.
10. Continue to notice what seems different as you go through the next few hours. You may be surprised. Essentially you are now using more of your body to gain extra vantage points of perception.
11. As an alternate to this exercise, you can try measuring your shifted awareness in the heartspace by using a work of art or another object. In class one example we use is a work of art by Salvador Dali. From the meditative state the students are able to add perceptions of the work of art that were unavailable to them prior to the meditation. While the expanded perceptions are highly personalized, they are likely to include added perceptions of colors, depth, tangible objects and action/intent.

Congratulations. That is all it is. But, this shift in awareness can give you access to information that you cannot attain when you keep your awareness locked in front of your eyeballs like a laser beam. We do this exercise every day. It is well worth the 30-seconds and can literally change your capacity to notice key signals in your daily work and home activities.

Taking Heartspace-Based Meditation on the Road

It makes perfect sense to use this 30-second meditation during your research-oriented day at work whether in the lab, office or classroom. But perhaps it has not occurred to you that this and other meditation tools can be equally useful on the road. I (RRD) was first introduced to the idea of taking meditation on the road for research insights by a very talented practitioner from New Jersey.

On the eve a large annual scientific conference in Baltimore, we attended a meditation practice event near the conference site. The leader not only gave us wonderful instruction and practice opportunities in heart-based meditation but also left me with homework. She told me to go to my national scientific conference, stay out of my left brain and remain in my heartspace just like where I had been at her meditative session. My first reaction was: Oh really! Then just as I gave it serious consideration, lots of thoughts ran through my head designed to raise doubts. After all, I might: a) wander into the Inner Harbor, 2) get lost in a Convention Center corridor and have to call my wife to come find me or c) be found trying to play center field in Orioles Park at Camden Yards (besides, I am better at first base). But I finally determined to focus on how I might best do this and also how I might notice the outcome of operating at my national scientific meeting in a state of heart-based meditation.

At the time the idea of attending a scientific conference essentially "out of my head" was a novel concept. In fact, prior to the most recent national conferences, my subspecialty had taken great care to develop grids of programming most relevant to our expertise so that we could fill up every minute of the five day-long conference attending talks that were the ones we logically should attend. After all, none of us would want to accidently wander into the wrong session and waste our time hearing about something unconnected to our own research. At least that had been my prior mindset and conference practice. This year's was about to be very different.

After taking care of some preliminary meeting tasks such a completing registration during the first morning, I set about to use the heart-based meditation for the full afternoon program. The meeting was organized as many national conferences with concurrent plenary lecture sessions based on specialized disciplinary themes organized in 3 hour blocks of multiple short lectures. With the Baltimore location, this meant that there were 6–8 concurrent sessions held in rooms that were laid out in a straight line down a central convention center corridor. Each room was the same size, the entrance to each was identical (or a mirror image) and the only distinguishing feature was the program theme itself (identified on a placard outside the entryway). This was perfect for using nonlinear tools to choose what specific session I would attend and what information I would hear during the afternoon.

A Personal on-the-road Experience

Here is how I (RRD) spent the afternoon. I picked up the program book but did not look at an electronic file of the program and ignored all session placards. I performed the 30-second heart-based meditation and walked to the corridor selecting the door that most drew my attention. I entered the session, stayed near the back of the room and heard the current talk. If the talk was near completion, I stayed for the next full talk. Once the talk was over, I exited the room, repeated the heart-based meditation, walked the corridor and again selected the door that most drew my attention at that time. Again I attended a full or nearly full lecture. I continued this process for the three hours of the conference afternoon sessions.

What did I notice? First, I noticed that I ended up in sessions I would never have selected based on the advertised program content. In fact, several of them I would have avoided at all costs. But what was amazing was that in every lecture or the following question and answer sessions, tidbits of information showed up that I really needed to hear. Even if a lecture seemed off topic, then the Q and A drove the discussion onto something with a novel idea that I could latch onto for my own specific application. In the end, it was the most insightful afternoon I have ever spent at any research conference, and it is continues to affect my own scholarly efforts. Was it really off-topic? Apparently not.

The lesson I learned was that while I might need some didactic, logically-identified information on the program at a scientific conference, I also had been missing golden opportunities to obtain information that could help me make quantum leaps in my thinking about my research. The novel ideas were in the seams of the logically-relevant programming, and I had been deliberately avoiding those seams because I had viewed those forums as irrelevant to my "universe."

Note that I am not recommending that you pick up a scientific conference program and deliberately set out for the most topically-irrelevant lectures you can identify in the program. That is not what I did. Instead, I simply found doors that interested me and caught my attention. I did nothing more. What was behind the doors was still a mystery until the "Ah-Ha" moment struck. In fact, in one or two of the sessions where the speaker had already begun the talk, I had trouble figuring out exactly what programmatic theme I had walked into. Another suggestion is do not judge the value of the session you might hear behind the door based on your immediate reaction. For a couple of the talks, it was not until I was travelling home that the novel idea associated with that session showed up. Be willing to have the useful insights continue to show up after the conference is over as the scholarly reward you so deserve for paying your conference registration fee.

THE HEARTSPACE-BASED MEDITATION ROAD EXERCISE

We recognize that not all scientific conferences are so conveniently organized as to have a single corridor with multiple identical doors as entrance-ways to concurrent scientific sessions. Therefore, variations on the personal door exercise may be more helpful to apply at one of your own upcoming scientific conferences. One example is described in the following:

1. Write the room numbers (without the session titles or topics) for the upcoming period (day or half-day) on identical-sized squares of paper and place them in a convenient container.
2. Perform the 30-sec heart-based meditation.
3. Draw a square with a session room number from the container. (Return the square you drew to the container so it remains a future option.)
4. Enter that session, notice what occurs and stay as long as you feel it is appropriate.
5. Then from the heart-space draw another room number and go to that room location.
6. As before, stay as long as it seems useful before repeating the process.
7. Notice what unexpected pieces of information you encounter and/or unexpected ways you may use routine information that has come into your awareness.

There are a myriad of other ways to vary this exercise such as with speaker names, session chair names, or a directional compass of some kind. The point is to go with what shows up from the meditative state and not from a linear, logical disciplinary or lab technique-driven basis.

TEN SIMILARITIES BETWEEN MINDFULNESS MEDITATION AND HEART-BASED MEDITATION

One of the things that struck us concerning two of the meditative approaches we discuss here is their overlapping similarities in many areas. Below is a check list of features that we find particularly useful for the scientist seeking new perspectives.

1. Exercises are designed to draw attention deeper into the body and away from a focus on the brain and head.
2. The exercises need not occupy much time. But daily or regular routine is encouraged.
3. When thoughts show up, asking questions around them can be useful.
4. There is no intention on outcome. The only intention is on the regular session and a possible shifted awareness.
5. The meditative exercises are portable and can be done virtually anywhere although some conditions may make them easier to accomplish.
6. A child-like playfulness and openness is encouraged.
7. The meditator is the observer and the observing is done without judgment or labeling.
8. The meditator can become a more accomplished observer.
9. Potential benefits are multi-faceted and can show up in health benefits, reduced stress, and increased creative problem solving skills.
10. The effects of these meditations can extend to those with whom you interact and large groups. In fact, they can be very effective in group settings (e.g., faculty and lab meetings, research forums, brainstorming sessions, strategic planning activities).

3. *Transcendental Meditation*

Transcendental meditation (TM) is thought to have arisen from the tradition of Vedantic Hinduism.[31] Many sources contributed to the modern-day form of TM as introduced by the Maharishi Mahesh Yogi.[32] He brought this meditative approach from India to the United States in the 1950s and from there it spread to many parts of the globe. Centers such as the Maharishi International University College of Natural Law in Washington, D.C. and the Maharishi University of Management in Fairfield, Iowa were formed, and the meditative tool eventually evolved to be taught as less oriented to a strict religious experience and more of a scientific consciousness technique.

The Maharishi Vedic Education Development Corporation via their Transcendental Meditation Program website recently described TM as follows: "The Transcendental Meditation technique allows your mind to settle inward beyond thought to experience the source of thought, the most silent and peaceful level of consciousness — your innermost Self"(http://www.maharishi.ca/). The meditative technique itself involves body-mind relaxation with a focus on what is known as a mantra. Rosenthal[33] describes the mantra or sound as a word derived from Vedic tradition that is given to the student by the teacher during TM training. Mantras are derived from a set of words or sounds and are selected specifically for the student based on certain criteria. Each student is asked not to reveal his or her mantra. Rosenthal indicates that the mantra alone is not enough. He calls it the "vehicle for effortless, inward movement of the mind."[33]

BEYOND AWARENESS WITH TM

As with other forms of meditation, the benefit of using TM for research creativity is associated with the different and/or expanded awareness. Altering one's awareness provides an opportunity to gain additional perspectives surrounding research problems. But as with other forms of meditations, TM appears to have potential health benefits as well. Two meta-analysis reports of studies examining TM and blood pressure concluded that measured but clinically meaningful reduction in blood pressure was associated with the practice of TM.[34, 35] TM has also been reported to improve both the quality of life and the functional capacity, as measured by a six-minute walk test, among African Americans with congestive heart failure.[36] One of the physiological effects associated with TM appears to be a reduction in cortisol levels as was seen in a study of postmenopausal women.[37]

SUMMARY

Meditation is an exceptionally important tool for bringing creativity to your research. By itself, meditation allows you to alter your vantage point and perceptional awareness such that you can access patterns of information that were previously beyond your mind's grasp. Additionally, meditation is an excellent adjunct to the other tools we cover in this book. We encourage you to find the personal form of mediation that works best for you and to try using mediation in combination with other tools such as play.

Chapter 16
The Neutral Observer: Letting Go of Drama, Ego, and Attachments

Where we have strong emotions, we're liable to fool ourselves.
— Carl Sagan[1]

Do you remember the last time you watched your favorite TV show, movie or favorite sports team in action. You were probably guessing what would happen next. You were silently, or perhaps loudly, cheering for certain characters to succeed or for one team to emerge victorious over another. In fact, it is a rare occurrence when people are drawn to watch a movie, a concert or a sporting event where they have absolutely no emotional interest in the progression of the event or the outcome. You might not make the effort to go unless you anticipated some emotional connection. It does happen. But it is more common among professional sports writers and movie critics who are paid to be analytical rather than emotional. Why is it we bring ourselves into full partisan spectator mode most readily when we are emotionally-entangled with the event we are observing?

Beyond the issues of science and technology, Tom Stevens[2] suggests that a key aid to being happy in life is to practice assuming the role of the dispassionate Hercule Poirot character. You should allow a part of your inner self to be the neutral observer where you are simply recording data, looking at the facts, and searching for the plain, unadulterated truth. He further suggests that a useful approach is to pretend you are an alien observer sent from another planet to observe and study Earthlings.[2] We have developed a brief exercise around Steven's suggestion:

THE ALIEN OBSERVER EXERCISE

1. Identify a two hour segment at your school or work when you can do this exercise. Ideally it should be a period where you will interact with people but not have a career-defining event (e.g., a final exam in a key course, a doctoral thesis defense, your annual department evaluation, etc.).

2. Pretend that you are an alien visiting this place for the first time. You have never seen the place or the people you will encounter over the next two hours. Your job is to move through the day without alerting the Earthlings that you are an alien. But more importantly, you must record what you notice. You need to do this without inserting your own interpretations or making snap judgments. Expert and careful observation is what is needed. Your colleagues back on the home planet are relying on your recorded observations and nothing more.

3. Record what you notice for the next two hours whether in the classroom or meetings.

4. What have you noticed now that you missed or discounted the last time you went through a similar two hour period in this same setting?

5. You have just assumed a useful neutral observer state. Keep in mind you can do this again when useful. You also have a calibration of what differed about your approach to noticing details in your surroundings.

THE NEUTRAL OBSERVER

One of the ideal states to achieve as a scientist is to play the role of neutral observer. The neutral observer may be passionate about the activity he or she is engaged in but needs to avoid being drawn into the emotional terrain of the surroundings. Dana Lightman[3] refers to this as "detachment" and advocates awareness of your perceptual position as an observer. To us this is the process of stepping to the side and observing a pattern of information without enmeshing yourself in it. Lightman[3] also uses the analogy of stepping outside yourself to achieve the observer position.

What are the characteristics of a neutral observer? As a neutral observer you are curious about what your research might produce. But there is absolutely no judgment, just the observation of new and useful information as you propose hypotheses, design and conduct experiments to test the hypotheses and collect, analyze and interpret the data as they shed light on the hypotheses as well as on alternative explanations. Creative insights, novel ideas, precise execution, and neutral evaluation are all helpful. You are much like Father Time watching the research process unfold. But the way forward and the exact outcome should remain undetermined and open to multiple possibilities much like Schroedinger's cat, which remains both alive and dead inside the unopened box.

To truly achieve this place of neutrality can be a significant challenge. This attitude is similar to non-judgmental openness we see in children. Children are often excited but are also perfectly willing to witness virtually any outcome. For children, the unexpected can be as exciting as the predictable. As adults, it can be exceedingly useful when we can let go of any need to control the path forward and the specific outcome we will reach.

DRAMA IS ALL AROUND YOU

An important distinction is the difference between being passionate for your work vs. running drama about your work. The former is a useful if not required component for a long-term research career. While intensity may ebb and flow over the course of a long career, the researcher will need to enjoy what he or she is doing. As with the example given for Barbara McClintock, it is useful to awaken in the morning literally burning to embark on the day's research effort. Confusing enthusiasm for the job with passion for outcome is the danger. It is very easy to power your life by what we would term, running drama. From 24-hour news networks to action computer games, online auctions, buy-it-now buttons and other entertainment vehicles, we can power our day by moving from one drama to the next. When this becomes our comfort zone, removing ourselves from drama can be not only difficult but also a scary proposition.

Rieger[4] in her monograph detailing the history of first radio then television soap operas points out that these can be found in virtually every country in the world. Instead of representing the occasionally-staged Greek tragedy of antiquity or a 19th century Victor Hugo play for occasional entertainment among Parisians, soap operas are an almost daily programming event. As Rieger describes, this societal phenomena is one of the few genres where weddings are not a happy event, but instead are a prelude to some doomed failure.[4]

Drama is one of the greatest and often underappreciated threats to creative research. By this we do not mean simply emotion. Unless you have Asimo on loan from Disneyland in your lab, you are working with humans and there will be emotion and probably some drama. Instead, we mean the powerful emotional programs that swamp our physiological systems and help to narrow our focus. These types of raw-emotional patterns can cause us to lose touch with parts of our body and shift us into auto-pilot for fight or flight mode. As a researcher, if you are in flight or fight mode, you are not likely to be that useful to the research program (unless you are researching the tensile strength of new suture materials). We all slip into drama and at times that is natural. But understanding and calibrating exactly what effect it can have on you as a researcher can help you avoid potential problems.

So much of our present-day media information outlets literally bombard us with drama. You do not have to be watching a daytime soap opera or prime time crime drama to find yourself intertwined in drama. Most advertising aims to ramp up your physiology before the commercial is over. The wireless and wired news is littered with the all-too-common, age-old themes of greed, lying, war, adultery, financial and health stressors, growing pains of children, personal tragedies, death and loss. Even when you think you are not experiencing these patterns of drama, if others around you are, you are likely caught up in it. It is exceptionally challenging to avoid drama, but it is also exceptionally useful to reduce it and to have some level of command over the extent to which drama can impact your work effort. If the old adage has been don't bring your work home then a revised version might be, don't take drama into your research space.

You can calibrate the effects of drama on your perceptions and neutral observer status in several ways. If you remember the Body as a Weathervane Exercise back in the chapter on the body, you had an opportunity to see the effects of reading a few news stories on both your physical body as well as your perceptional awareness. It can be quite ….. dramatic. Another exercise we use in the class and workshops is to provide a dramatic context to an otherwise purely informational brief abstract or statement. Participants read the same statement with: (1) no visual cue, (2) with the picture of an atom bomb exploding under the words, and (3) with a cute puppy picture placed under the words. The nature of the visual cues including a vision of war and destruction can affect how you react to the words you are reading. This may be easily understandable, but we often underestimate the pervasive nature of drama around us and how it can affect our day.

RELINQUISHING EGO AND YOUR PERCEPTIONS OF A SINGLE WAY FORWARD

At first glance, what we are about to describe may seem to be counterintuitive as something useful for research success. After all, you need to believe in your own work, the direction that it leads you and the value of overall contributions you can make through research and technological innovation. Confidence is crucial. If you do not believe in yourself and your proposed research, who will? But it is important to distinguish between confidence in your capabilities, vision and research tools vs. ego-driven paths of advancement. About the time you have decided that you not only know where the research is going to take you but also know precisely the path you must take to arrive at the beneficial outcome, it is about the time to start worrying. Like the mutual fund financial advertisements are required to state, past performance does not predict future performance. This also occurs in the paradigm-determined pathways that lead to useful research outcomes. Unless you are taking the final step and are about to land in the outcome, you are probably better off not to judge the way forward based on a linear extrapolation from where you have traveled. Our egos can push this extrapolation incessantly. By leaving open the specificity of the path forward, as occurs when you have neutral observer status, you can actually gain a broader perspective of the paradigm-driven path you are proposing.

I (RRD) recently had a personal experience with this very dichotomy. The beneficial outcome of reduced prevalence of immune dysfunction-related chronic diseases can be connected, at least in part, with better avoidance of adverse early-life environmental exposures. Better identification of problematic exposures would help reduce this disease risk. For the past few years and along with several colleagues, I have suggested that to achieve this desired outcome, we are likely to need disease-relevant and age-relevant immune safety evaluation of drugs and chemicals. This is called developmental immunotoxicity (DIT) testing. In other words, a different approach to required safety testing is needed. This recommendation has been previously published as a way forward for better health protection during childhood.[5-7] That would seem to be the logical solution to the current paucity of relevant developmental safety information. But is that the only possible way forward to the desired outcome? Are there any other solutions including those that are very different from what I have suggested?

A different path to a solution would be viable if: (1) We could assume that currently collected data are underutilized and more information could be squeezed from existing data and (2) We knew how to better and/or differently use these data in novel ways to reach scientifically-sound conclusions regarding age-specific health risks for chronic diseases. It should be noted this is a very different approach as it is completely counter to my call for the collection of different data. The focus is then on identifying previously unrecognized ways to gain insights from what are otherwise generally limited, non-functional safety data.

I set about to investigate this possibility recognizing that, in one sense, such a path forward would be a stark alternative to the path I had been forging. The way I devised a strategy for more extensive use of existing nonfunctional data in an unchallenged system was to examine the fractal dimension aspect of the immune system. In this context, I presented the concept that scaling properties could allow new information to be obtained from existing static safety data. The paper on fractals and the immune system was published in 2011.[8] As of yet, no one has shown that using fractal dimensionality of immune safety data would extend its utility and substitute for the new DIT testing data that I have suggested needs to be collected. However, by spending the time to consider and then publish an alternative strategy to my primary paradigm, I was able to define the problem better, offer a more complete range of plausible solutions and, hopefully, attract new investigators and investigations into this area of children's health protection.

As a neutral observer researcher, the importance is to see the outcome of better protection of children's health. The precise route used to that end is really comparatively less important. Researchers do not always have the luxury of spending the time required to publish papers surrounding alternative paradigms. But a few minutes practicing neutral observer status can be useful when it comes to broadened perspectives and enhanced scientific discussions of ongoing research activities.

ATTACHMENTS

One of the opportunities for personal discovery for the "independent scientist" is to examine what we call attachments. Attachments are essentially the hooks that lock us in place and prevent us from moving forward. Without a complete examination of your potential attachments, the "independent" part of independent scientist may be more theoretical than actual. Usually we are not aware of these impediments. Because they are numerous, generally hidden, and often well-entrenched, attachments are not shed as easily as your winter coat. In *Inner Excellence*,[9] Jim Murphy defines eight attachments that are important in locking us into one place: (1) how others see us, (2) our money and possessions, (3) what we want, (4) comfort, (5) our past, (6) resistance to change, (7) expectations, and (8) ego.

Enmeshed in attachments is the issue of how we view failure. Murphy[9] points out that we tend to personalize failure. We tell ourselves the story that failure in a single effort is actually a poor performance and possibly even failure as a person.[9] But one of the lessons we learned in the previously discussed, Martha Beck-inspired, write-your history exercise is that apparent bad events can be the keys to our later-life successes. A failed experiment, an unfunded grant proposal, a rejected paper may be the event that sets you on a more useful scientific path. What if you must fail in one effort to succeed in your most significant accomplishment ever? Can you withstand the risk of a failure if that is the way forward? Would you brave that possibility? These types of questions provide an opportunity for you to calibrate your willingness to: (1) break attachments, (2) leave your comfort zone, (3) shift expectations, (4) set aside ego, and (5) welcome change.

I (RRD) can relay one recent example that supports the idea that failure can be different than it seems and sometimes is a blessing in disguise. It can represent one door closing and another opening. Shortly before *Science Sifting* and the related university course came to life, I had another opportunity to prepare a collaborative lay-audience book on a topic much closer to my actual scientific training and prior work. Preparing the book proposal involved a substantial writing effort that, after months of writing and leg-work, went nowhere. At the time it was a significant disappointment given the time and labor involved. However, fast forward two months and it became clear that if that book project had moved forward, the current effort in enhanced creativity for scientists might not have seen the light of day. That more traditional, collaborative project needed to fail for the innovation initiative to have the chance it deserved. Sometimes you need the decks cleared to make way for your future success.

SHIFT TO NEUTRAL

The many workshops, seminars, and training weekends we have taken over the years have included an often-encountered recommendation to "shift to neutral" as a personal exercise. This can be done daily, weekly, or as needed and take one of many different forms (e.g., calibration within your own body, stepping out of the drama and attachment state, giving yourself an alternate focal point). Some of the exercises involve visualization, others meditation and still others physical body movement.

Personally, I (RRD) use a combination of heart-based meditation with a visualized adjustment in my own body. It is much like watching the indicator arrow on my body's drama scale shift back to the zero position. I try to do this on a daily basis usually when I am at a stoplight on my way to the university. It is that quick and that simple, but it works for me. By the time I arrive at work, I am prepared for whatever issues may show up during the day. You might do this by using your favorite music as in the body chapter or through another of the many tools we discuss or you encounter elsewhere. However it works for you, shifting to neutral can be a huge and freeing change that opens your personal creative space.

SUMMARY

Operating as a neutral observer is one of the more useful actions you can take to enhance creativity and innovation in your science. You may think that you always operate as a neutral observer. After all, you were likely trained for many years to be a neutral observer of research and to exclude judgments and biases in your work. The reality is that you are swimming in a sea of drama that bombards you even in public places like airport terminals. That daily bombardment is something you were probably not trained to manage. Intentionally removing yourself from that drama and its impact on your perceptional awareness can significantly increase your effectiveness for unbounded creative thought. Stepping out of drama and the attachments that lock you into a narrow status quo perspective is useful. This may require a loss of both ego and the illusion of control, but the benefits are great. One strategy is to use the tools we and others have discussed for correcting to neutral on a regular basis. Even reminding yourself of the benefits of correcting to neutral can make you more aware of the daily drama you are studiously avoiding.

Chapter 17
Play Like You Mean It

My occupation is a simple one. I play with microbes.
— Alexander Fleming[1]

If you want creative workers, give them enough time to play.
— John Cleese[2]

When you're a child, something as simple as a tree doesn't make
sense. You see it in the distance and it looks small, but as you go
closer, it seems to grow — you haven't got a handle on the rules
when you're a child. We think we understand the rules when we
become adults but what we really experienced is a narrowing
of the imagination. — David Lynch, filmmaker[3]

Diane Ackerman, author of *Deep Play*,[4] describes human play as "a sanctuary of the mind where one is exempt from life's customs, methods and decrees."[4] Think of how freeing that is to opt out of the very things that limit you. Play is one of the key strategies for getting outside of the box. That is one of the reasons that the Stanford University Institute of Design offers a university course titled "From Play to Innovation" linked with the textbook *Make Space*[5] by Scott Doorley and Scott Witthoft. Of course, seeing new patterns directed toward innovative new models of design is really not different from seeing: 1) new patterns of chemical interactions for bacteriophages, 2) new patterns of nanostructures, 3) new approaches to the design of medical devices, 4) new approaches to emergency preparedness, or 5) new patterns for food agriculture sustainability. There is no reason that play would be restricted in usefulness to one research disciplinary area. The larger question is, why do we have to reteach young adults and aging researchers how to play like they are kids? It seems likely that we taught it out of them somewhere along the way to graduate school. It is challenging, yet important, to emphasize just how important the role of play can be for inviting novel patterns to appear in your awareness as you pursue complex research problems. Even in the aged, there is the concept that child-like play provides not only enhanced creativity but also a form of compensation for age-related functional loss.[6] Playing with toys has a long tradition particularly when you consider that one of the oldest toys from Europe is a clay horse that dates to 9th-10th century B.C.[7]

To emphasize this, we suggest you, at least momentarily, exchange the traditional labels given to techniques and aids of innovation as they are routinely discussed in other books for an alternative. Most commonly these aids are called tools in a tool box or tool kits, or buckets. Instead, you can just as easily refer to them as toys in your toy box. Why should you do this? You may even find this to be irritating or disturbing. Will you take these aids seriously now? We hope so. The reason we make this suggestion is that tools are an adult term. Most adults know they need to use the right tool for the task (e.g., a hammer, wrench or pliers). If you need to do plumbing work, but can only find a hammer you are unlikely to be successful. You do not have the right tool for the job. But the child is much more flexible. A child will take whatever prop is available and make it right for today's game. We are unlikely to devote an entire chapter to Green Hornet's lantern as an aid for research innovation (at least not in this book), but the aids we identify can become toys sitting in your personal toy box. When you use them, it is really play, not work. With at least one toy in hand, you are ready for today's game. Research-oriented work is often prescribed, serious and bounded, but play is open to whatever shows up.

For many adults, play has been stereotyped into something that is far different from its possibilities. As described by Martha Beck,[8] when most adults are asked to pick out images of adults at play, they choose photos of people lying in a hammock, or sipping beverages on a beach. These sedentary forms of "play" are actually sleep-deprived cries for rest and do little to enhance creativity. But it is hardly the full range of activities available. For adults and probably many serious scientists, play has become inactive and deliberately boring. We think we are clearing our mind by lying in the hammock and expect that wonderfully creative ideas will rush in. That could happen. But it is far more likely to occur through what Martha Beck also calls "deep play."[8]

Deep play is more what you see with children. In fact, if you have not done it recently, watch young children at play. Such play challenges and activates rather than dulls. A child becomes Superman, the Lone Ranger, a Power Ranger, Wolverine, Transformer, or Xena, Buffy, Rogue, Wonder Woman, Cat Woman, Jean Gray or Storm. There are villages, if not whole civilizations to rescue, and in the child's mind, he or she is that beyond-normal-human archetype. In other cases, a child will spin a top until a new personal record is achieved, take the slinky down even more stairs than before, dance in circles until standing is no longer an option, play dolls or house adding new features each time the game is played, or pretend to be different animals making the sounds more ferocious or distinctive and imagining animal movements around a storyline.

Play is the thing and this type of play that seems merely cute to an adult is really transformative to a child. In fact, this play would be similar to adults immersing themselves in Robert Barker's panoramas, Kevin Brown's mock villages, or playing in a set with a garage band using Richard Feynman's bongo drums, Barbara McClintock's banjo, or Richard Flavell's guitar (see the next chapter). You are immersed in these experiences and your prior, more limited reality has been shoved to the side. From here, you are open for connecting new dots and seeing new patterns of information.

Perhaps we should retitle the book: "Ziplining to a Nobel Prize." It is not as wild an idea as you may think. Here is what Linda Stone reported in her 2010 article *Finding Ourselves Through Play*[9] concerning her interviews with Nobel laureates in Sweden. "When these men talked about their work in the lab today and their childhood play patterns, it was the same conversation. They played passionately as children and the emergent questions and interests they had as children were still central in their work, albeit more evolved."[9]

This type of deep play as described by Martha Beck and reported on by Linda Stone takes you outside of your left brain and outside of your box. It takes you outside of the mundane world easily taken for granted and quickly transports you into more creative spaces. Deep play can be used to disconnect from what is called conventional "wisdom," which, from my

(RRD) own experience, can be an exceptionally useful disconnection. Once you do that, it then becomes your game, your rules. Sound familiar? That is because when you were a child engaged in play, it was likely to have been your game, your rules. Once you start to tweak the rules and operate outside of conventional "wisdom," your information filters begin to fall apart.

This is what allows you to see new patterns that were not available to your awareness before. At this point you may be asking what a new rule set might look like. For me, this is a form of deep play. I (RRD) put a new rule set in place a few years ago that really obliterates conventional wisdom. I call it "less work, more results." Of course conventional wisdom says that work and results are generally linearly related. But that is not a useful rule. It would guide you to think that research innovation and success only comes by working yourself more and more until you are showered with accolades all given posthumously. With "less work, more results," I essentially moved into nonlinearity in my academic work. I do not mean to imply that this book was written by some unseen hand while I spent months at a beach. Rather, it is just that I do not have to spend more time at or in work to see an increase in satisfying results. Once you break out of linearity and enter nonlinear spaces via deep play, other restrictive linear relationships dissolve. Deep play is one of the most useful toys in your toy box for bringing more creative thinking and innovation to your research career. Are there examples of useful play connected to enhanced research creativity? Absolutely. Here are five examples of the play of brilliant scientific innovators.

Alexander Fleming

There is evidence to suggest that Alexander Fleming's mantra about playing with microbes (our opening quotation for this chapter) was a very accurate description of what he pursued at work. Fleming played with microbes to such an extent that he created paintings using them in place of paint.[10] This was consistent with Sir Almroth Wright's expressed view that Fleming treated research much like a game (described in Root-Bernstein and Root-Bernstein[11]). It is easy to see the childlike openness that is so useful for discovery in these characterizations.

Albert Einstein

In our next chapter, *Striking a Chord of Creativity*, we discuss the importance of music to Albert Einstein. He was an avid and quite proficient musician playing both the violin and the piano. Einstein used this diversion to enhance his abilities to see new information in theoretical physics. It may or may not seem ironic that music and string theory are purported to have overlaps as discussed in the book *Einstein's Violin*[12] by Joseph Eger.

Brian Foster[13] in his article titled *Einstein and his Love of Music* quotes Einstein as saying that: "Life without playing music is inconceivable for me...I get most joy in life out of music."[13] Music was not the only playful activity that inspired Einstein's creativity. Einstein was an avid sailboat sailor. This seems unlikely and all the more remarkable since he could not swim. He spent virtually every summer of his life in the U.S. at a vacation home near the water so he could sail.[14]

Leonardo da Vinci

Leonardo da Vinci was notorious for filling his handwritten notebooks with drawings of everything from human anatomy to art sketches to machine models. He was a remarkably keen observer who was forever drawing what he noticed. In fact, he is estimated to have produced 52,000 drawings in his 67 years of life.[15] Some of these hand-drawn models he built and operated; others may never have left the drawing board. In 2011, the Da Vinci Museum of Leonardo Da Vinci in Florence mounted a multi-year, world-wide, touring exhibition titled "Da Vinci Machines Exhibition" bringing 60 of da Vinci's models to life.[16] In general his machines have been subdivided into five categories: earth, water, air, fire and mechanisms.[17] Among his mechanism-cateogry machines are a humanoid robot with gears that may have permitted a drumming sound.[15]

In many of the earliest discovered notebooks, the models seemed largely purposeful and usually with a practical single purpose in mind. But with the discovery of two "lost" notebooks amounting to approximately 700 pages in Madrid in the mid-1960s, it has become clear that some of these models were well beyond single-use solutions. In fact, Anna Maria Brizio[18] emphasizes that da Vinci made no real distinction between

artistic and non-artistic drawing or between his scientific or artistic activities. To him they were one and the same.

This is the key point in considering play as an opening for scientific
insight. Da Vinci spent his time playing with the information he noticed from
an amazing array of sources that surrounded him. He was less concerned
about categorizing or pigeonholing his information. He moved seamlessly
between art and science simply working with the patterns of information. If
you watch children at play, their approach is much the same. Adult's catego-
rize, set boundaries, and label, children do not until they are taught to do so.

Richard Feynman

Nobel laureate Richard Feynman had a notorious sense of humor and man-
aged to insert some aspect of play into nearly everything he pursued. For
example, at an early stage of his career, he was sequestered in virtual isolation with a handful of scientists at Los Alamos working on one of the most
serious scientific projects in U.S. history (The Manhattan Project's development of the Atom bomb). In response to sheer boredom, Feynman amused
himself by overcoming virtually every security system in the facility. Not
only did he successfully break into most of the systems at Los Alamos, but
he extended this amusement to the safes at the Oak Ridge National
Laboratories during a couple of visits to those facilities.[19] It became clear
that your safe was not safe if Feynman had no other games to play.

Andre Geim and Konstantin Novoselov (2010 Shared
Nobel Prize in Physics)

As described in the official press release from the Royal Swedish Academy of
Sciences (RSAS),[20] Geim and Novoselov won the prize by "Using regular
adhesive tape... to obtain a flake of carbon with a thickness of just one atom.
This at a time when many believed it was impossible for such thin crystalline
materials to be stable."[20] Adhesive tape! Who would have thought? In describing the awardees, the press release continued by saying that "playfulness is
one of their hallmarks, one always learns something in the process ....."[20] This
laboratory group thrived on regular play in the lab. Through an activity that
arose not by design but rather spontaneously during their play, they made
what is already seen as a monumental scientific breakthrough.

Benjamin Franklin

Benjamin Franklin had several outlets of play that were an important part of his creative life. Most of his play-related activities, outside the realm of his printing job, spawned additional inventions. For example, Franklin was an avid swimmer. To aide this activity he invented swim fins that gave him extra speed in his youth and kept him with enough energy for swimming into old age.

Franklin was playing checkers[21] (then known as draftboard) by 1726. He took up chess[21] at least by 1733 and was among the first chess players in the colonies. In fact, Franklin used chess for diplomatic purposes as demonstrated by his play with Lady Howe in London on several occasions and the eventual meeting with her brother, the admiral, who would later lead an army against the colonies. Chess was the entry to a final serious attempt at negotiations (drafted terms of agreement).[21]

Franklin played for much of his life and even wrote an essay on the subject saying that "Chess is not merely an idle amusement. Several very valuable qualities of the mind, useful in the course of human life, are to be acquired or strengthened by it, so as to become habits, ready on all occasions."[22] From his descriptions of chess you can see the role that it played in broadening his relational perceptions of nature. He wrote: "Circumspection, which surveys the whole chessboard, or scene of action; the relations of the several pieces and situations, the dangers they are respectively exposed to, the several possibilities of their aiding each other, the probabilities that the adversary may make this or that move, and attack this or the other piece, and what different means can be used to avoid his stroke, or turn its consequences against him."[22] Benjamin Franklin was inducted into the U.S. Chess Hall of Fame in 1999.

Franklin's mentally challenging "games" were not restricted to chess. Unknown to many, Benjamin Franklin was among the most important colonial mathematicians. He used complex mathematical "games" such as magic squares and magic circles to pursue this interest.[23] The exercises probably contributed to his capacities for weather forecasting.

In addition to swimming, board games and mathematical puzzles, Franklin loved music. His musical activities supported his play and his inventiveness and are discussed in detail in our *Striking a Chord of Creativity* chapter.

By now you are hopefully appreciating that research is serious business until it's not. Play breaks the attachments that hold you back from seeing what you need to see. The opportunity to bring a playful openness to your research roadblocks is your all-access pass for seeing new and useful patterns of information.

Barbara McClintock

In closing this section of play, we would point to one of Cornell's most famous women scientists, Barbara McClintock, geneticist, botanist. Through her research on maize, Dr. McClintock discovered genetic transposable elements that explained the way in which genes can be rearranged and regulated. She did this by noticing color patterns in the kernels of maize. She observed patterns that should not have occurred based on known genetics; patterns that were anomalies. She did not discount these anomalies. As is discussed in other parts of this book, her discoveries went largely unnoticed for 30 years.

What is less well known is that McClintock had a marvelous sense of humor that she took with her everywhere she went even into seemingly serious or dire circumstances. As described by McClintock's colleague, Dr. Evelyn Witkin, McClintock's laugh could rock buildings with its force.[24] She once noted an error on a White House invitation register (to an event sponsored by Henry Kissinger). McClintock jotted a quote in the log margins that was the best wit of Rodney Dangerfield. How appropriate that McClintock would choose Dangerfield's humor at such an august event. You may recall that Dangerfield's trademark comic line could have been an apt description of a portion of McClintock's career, "I tell you, I just don't get no respect."[25] Finally, when in South America fighting a serious infection and receiving condolences from friends, McClintock commented that the infection was so interesting.[26] Humor can get serious scientists through serious roadblocks.

Beyond becoming a five year old again, what can you do to bring more play into your sphere of research? Here are three exercises in a row that provide varying degrees of challenges. These can provide you with wiggle room for more play in your everyday scientific pursuits.

THE MORE WORK OR MORE PLAY EXERCISE

This is a hypothetical situation unless you happen to be heading to Anaheim right now for a scientific conference. But by engaging this situation, you can calibrate the amount of wiggle room you have for blending science and play.

1. You are attending a large science conference at the Anaheim Convention Center. Your talks are over, you have already had three solid days of meetings and none of your students or closest friends are on today's program. But there are more sessions on the program. Disneyland is right across the road and you have not been there since you were a kid.
2. Ask yourself where you are more likely to get a creative idea: visiting Fantasyland, the Tiki House and Pirates of the Caribbean or attending the last few lectures of your science conference? Now you can go to Disneyland without guilt (and please tell Asimo hello for us).

Bringing play directly into the work space can be very rewarding.

THE OFFICE AS A WORK OR PLAY SPACE EXERCISE

1. Pick out a toy (if you don't have one handy you might have to invest in one or borrow one from a child).
2. Bring the toy to your work. What do we mean by a toy? It could be anything that is clearly recognizable as a plaything by a majority of people walking into your office area. I (RRD) will use my special Neutralizer from the movie *Men in Black*. It is great for erasing memories of outdated paradigms.
3. Place the toy on your desk and leave it there for at least a week. [Good example — Penelope Garcia on the TV show Criminal Minds.]
4. Pick up the toy and play with it briefly each time you enter and leave the office and notice what happens. You have effectively turned your work space into play space. The insights you get, the productivity during the week, the interactions with your colleagues are likely to be different. It is now as much a game as serious, high-level work. Additionally, it is your game.

THE MY RESEARCH IS SERIOUS EXERCISE

This exercise is designed to take the seriousness and stress out of work and to open you up for broader awareness of information. While it is deliberately absurd, it can be very useful.

1. Find the latest scientific abstract, project summary, or grant proposal specific aims you have written.
2. Using your computer, find an image of the 1980s television character ALF (from the TV show named ALF), which stands for "alien life form." If you have never seen ALF before, check out YouTube or other videos of old episodes and prepare to be amazed.
3. Sit facing the image of ALF as if he were the most important cosmic being (like the Queen of England or members of the study section where your grant proposal is likely to go) you have ever addressed in your career. The future of your research path could hinge on your communication with ALF.
4. Now read the description of your research to ALF. How do you view your relationship to this work after this brief extraterrestrial experience? What might the TV character ALF have to say about it?
5. Notice how you approach your scientific work or studies in the next 48 hours.

This last exercise is not intended to diminish the importance of your research. Quite the contrary. By presenting your research in what amounts to an extraterrestrial comedy club, you have momentarily made your research a game. Your research is now open to more possibilities. You are playing with your research like Alexander Fleming. It is likely you will notice reduced stress around deadlines and tasks. You are likely to become more aware of information that is useful for your scientific efforts. By letting go of the seriousness, you unbound your relationship to your research and, ironically, are far more likely to make a seriously-important observation.

SUMMARY

Play provides you with a profoundly transformative opportunity to enhance your scientific creativity. It is also one of the most readily available aids. Introducing play into your science can be as simple as putting toys on your desk and using them each day or letting ALF critique your next grant proposal or report. Many of the most innovative scientists of our time used play and recognized its importance to their science. You can access information during play that is beyond anything you can reach by continuing to come up against your research roadblock. Look around your office or lab and think about how you might play with something already there. The next time you are in a department store, go toy shopping. Play pays dividends in ways that are difficult to imagine.

Chapter 18

Striking a Chord of Creativity: The Art of Science

I often think in music. — Albert Einstein[1]

The greatest value of a picture is when it forces us to notice what we never expected to see. — John W. Tukey[2]

USING LATERAL PATTERNS TO CONNECT MORE DOTS

Music and art that cause you to see in images bring your awareness to patterns of information outside of your science perceptions. Regularly accessing lateral patterns appears to have a direct linkage to thinking creatively in science. This is becoming more widely appreciated as is the view that high levels of creativity in the arts and sciences are really not as distinct neurologically as had previously been thought.[3] We view this as habitually seeing lateral patterns. Along with vantage points, deep play and creative spaces, pursuing activities for perceiving lateral patterns is another toy in your toy box for scientific innovation. If you have passion for music or the arts, you are well on your way to increased scientific creativity.

According to Ken Robinson[4] in *Out of Our Mind*, we get creative insights through nonlinear processes when we see connections and similarities between things that we hadn't noticed before. Creative thinking depends greatly on divergent or lateral thinking and especially on thinking in metaphors or seeing analogies. We covered the use of metaphors in a prior chapter and both metaphors and analogies have been discussed by Roberta Ness[5] in *Innovation Generation*. These associations help you to connect with patterns. In addition to the word-based varieties there are audio-based, image-based, and body-sense or kinesthetic-based analogies as well. But perhaps the best associative forms of all for making nonlinear connections have been saved for last: music and various art forms.

MUSIC

Music has long been known not only to evoke specific brain patterns but also to have mathematics as a basic foundation. In fact, music can have a profound effect on patterns of thinking, behavior, mood, and various perceptions including visual perceptions. An area of study called cymatics is transforming sounds, and in particularly music, into complex 2-D and 3-D visual wave patterns.[6,7] There have been many technologies associated with music and visual perception over the years, but one like cymatics appears promising for revealing precisely how music is connected to very specific patterns of information that are visually available. Investigators like Joshua Leeds[8] have examined the potential for cymatics and other music-wave approaches to enhance not only performance but also health.

Another important aspect of music and vision involves color. As described by William Morris in his article "The Dream of Color Music And Machines That Made it Possible,"[9] ideas linking music and colors go back to Greek philosophers like Aristotle and Pythagoras. An optical ocular harpsichord was invented in the early 18th century, and the composer Telemann wrote music specifically for the instrument. Hungarian pianist and composer, Alexander Laszlo, wrote a theoretical work "Die Farblichtmusik"[10] (Color-Light Music) in 1925 and then toured Europe performing music with his own designed color organ equipped with switches for colored spotlights and slide projections on the stage above his organ.

Music has such a significant impact on mood and behavior that it has been used to calm, agitate or drive away crowds. From subways to bus stations, shopping malls to airports, the music is usually intentional and purposeful. In a recent Library of Congress podcast,[11] Dr. Jacqueline Helfgott of Seattle University explained how classical music is used by law enforcement and other institutions as social control to reduce the risk of crime. The lecture was in the series "Music and the Brain" and the podcast was titled "Halt or I'll Play Vivaldi."[11] Think of the fun you can have with possible alternative subtitles: "Can You Handle Handel?" "Back for More Bach?" "Tell a Man to go With Telemann."

THE SHOPPER'S DELIGHT EXERCISE

Pretend you are the store manager of a Walmart Superstore, and it is midnight on Black Friday. You are picking background music to be piped into the store and outside areas of the store preceding, during, and after the 4AM store opening. Do you pick from songs on list A or list B?

A) Steve Aoki in "Wake Up Call" (DatSik Remix)
    Screaming Jay Hawkins singing "I Put a Spell on You"
    Weird Al Yankovic performing "Polka Face" (If you haven't heard this... consider yourself warned).
    Semisonic singing "Closing Time" (just sends the wrong message)

B) Handel's Water Music
    The Platters singing "Only You" (aka. you know you want this item)
    Britney Spears singing "Baby One More Time" (aka. make another purchase).
    Elvis singing 'It's Now or Never' (aka. buy it now there is no second chance)

I suspect you would opt for songs on the B List as they are likely to be more in tune with a focus on pleasant shopping rather than an irritating distraction of mixed messages.

Not surprisingly, animals are also affected by music. In Cornell University research, noted animal physiologist, Ari Van Tienhoven, and colleagues demonstrated that piping Vivaldi's Four Seasons into chicken houses periodically at moderate volumes improved weight gain in broilers.[12] More recent investigators have shown piping classical music into chicken houses will reduce stress in the animals.[13]

Have you ever watched a favorite movie with and then without its sound track? It is an exercise well worth trying. Take your favorite movie saga, perhaps Lord of the Rings or Harry Potter films. Take away the music and it all falls apart.

THE MUSIC APPRECIATION EXERCISE

This is one exercise you may already have done, maybe often. But here we will do it with a specific purpose in mind: you creating a "reality" via music.

You can take your favorite stage (or staged) production with music such as a Broadway musical or an opera. Alternatively, you can use your favorite contemporary music video.

1. Play only the music and close your eyes. Can you see the action and images of the singing and body movements timed to the music? You probably nearly recreated the original integrated production in your mind.
2. Now, play only the music again. Try to envision a different production. You are the director but can you envision something different: different about the set, background, clothes of the singers, their interactions, the timing of their interactions, stage effects such as lighting, smoke, the props including both the ones used by the singers and the ones sitting in the background? How does the integrated production change?
3. If you have time, try it a third time aiming for making even more significant changes. Need ideas? Can you substitute the performers? You could start with easy substitutions: perhaps Shania Twain substituting for Bonnie Raitt? But then try more interesting possibilities: L.L. Cool J in a guest spot with ZZ Top or Kiss, Britany Spears singing a Shakira song or Justin Bieber substituting for Mick Jaggar with the Rolling Stones.

By the way, should you attempt to try this exercise using Wagner's Ring Cycle operas, be sure you have at least 16 hours to spare.

A recent example of integrated science, mathematics, and music can be found in the film, *Inception*, a movie about different consciousness levels, dreams within dreams. If you have not had a chance to hear it, we highly recommend the soundtrack composed by Hans Zimmer. Music is integral to the movie's plot line since musical cues are used to tell the characters which "dream" level" or reality they are in. It is their only signpost. The sound track has raised some controversy since a leitmotif type extract is based on a nugget from a previous song by Edith Piaf. But what Zimmer does is to mathematically manipulate the nugget and present it in almost unrecognizable variations that mimic the shifts in dream levels within the story.[14] Watch *Inception* without the music and you are not seeing the same movie. In contrast, listen to the sound track without the movie scenes, and you may well experience wobbles in your perceptions. Somewhere inside you, you understand the mathematical manipulations of the music.

THE MUSICAL TUNING ISSUE

One of the more intriguing and controversial debates about music and creativity does not concern rhythm, volume, category of music, instrumentation or active or passive music participation. Instead it concerns tuning. That is, before you even play a piece of music, what is your frame of reference for the hertz frequency that represents a given note? The importance of frame-of-reference is something we emphasize throughout the book. Here, the herzt tuning is your foundation for any and all of music. While different tunings have been used throughout history, the past few centuries the most commonly used standard was the A note pitched at 432 hertz (called A432). Curiously, during the 20th century, advocacy for a change from A432 to A440 herzt arose and in 1953 the International Standard Committee officially made this change. Now you will be hard pressed to hear any music played with A432 tuning. Music simply sounds different when pitched to A440 than A432 (also called the Verdi A) and the question is, does that difference affect behavior, emotion and possibly, creativity? Unfortunately, the issue has yet to be resolved through scientific consensus. If you can find A432 vs. A440 music and/or tuning forks, you may want to test the differential effects tuning has on you. After all, it is your potential for creativity this book hopes to aid.

MUSIC AND CREATIVE INNOVATORS

Galileo

The science-music connection is ancient and significant. For example, Galileo was born into a professionally-musical home and learned to play both keyboard and lute. At one point he considered becoming a professional musician. His father, Vincento,[15] performed experiments on science-math relationships in music and published expressions of consonance and dissonance as mathematical formulae. In fact, there is some reason to believe that Galileo, via the Medici family connection, may have influenced the invention of the piano by Cristofori. Galileo's experience of early music making, mathematical relationship to music and mechanics were likely important to his perception of patterns linked to his later scientific discoveries.[16]

Albert Einstein

Einstein was an avid and proficient musician playing both the violin and the piano. Brian Foster[17] in his article titled "Einstein and his Love of Music" quotes him saying: "I live my daydreams in music…. I see my life in terms of music…."[17]

For Einstein it was Mozart's sonatas for violin that brought him unimagined joy. His wife, Else, described Einstein's interplay between joyous play and work by relating that "He also plays the piano. Music helps him when he is thinking about his theories. He goes to his study, comes back, strikes a few chords on the piano, jots something down, returns to his study"[17]

In his book, *Einstein: His Life and Universe*,[18] Walter Isaacson tells how Einstein's friend described that "He would often play his violin in his kitchen late at night, improvising melodies while he pondered complicated problems….Then suddenly in the middle of playing he would get excited and announce, I've got it! As if by inspiration the answer to the problem would have come to him in the midst of music."[18]

Einstein himself described his improvisational strategy on the violin as follows: "First I improvise, and if that doesn't help I seek solace in Mozart. But when I am improvising and it appears that something may come of it, I require the clear constructions of Bach in order to follow through."[19]

So Einstein was able to use playing the music he so loved and improvisation to enhance his abilities to see scientific solutions.

Leonardo da Vinci

Leonardo da Vinci was not only a singer and musician but he was of sufficient talent that he was asked to perform at events of influential people. He also composed music, which is extant. But as described by Emanuel

Winternitz[20] of the Metropolitan Museum of Art, there is an even greater connection to music. The comparatively recent discovery of additional notebooks da Vinci left behind suggests that he designed several different types of unusual musical instruments. One of these, the "viola organista" was recently built and put on display. It has been described as the Renaissance version of a one-man band. The reason this instrument was such as breakthough innovation in 1488 was that it allowed the player to strike chords. During da Vinci's time most keyboard instruments could only play a single note at a time. Plucking stings of a harpsichord could be done quickly but still produced no chords. So da Vinci's harpsichord-like invention would provide more complex musical patterns for the listener. Apparently, da Vinci was already thinking about moog synthesizers when it came to music.

Benjamin Franklin

Franklin loved music. He played the violin, harp and guitar. But his connection to music and instruments did not end there. The first recorded American military band was in 1756 connected to an artillery unit under the command of Col. Benjamin Franklin. Benjamin Franklin got additional notoriety for another music-connected invention. Like da Vinci, Franklin invented a new instrument called the glass armonica, which debuted in 1762.[21] Both Beethoven and Mozart wrote music for Franklin's new musical invention. Beyond inventing and playing instruments, Franklin designed a four-sided musical stand. An extant example that was sold with other Franklin family furniture is made of walnut and attributed to William Savory. The stand allowed a quartet of musicians to sit around and easily play. In keeping with the newly-designed quartet-friendly stand, Franklin is thought to have composed one or two string quartets while in France.[22] The two quartets with Franklin's name on them were discovered and published in 1946. They employ specially pre-tuned instruments such that open strings (bowing, but no fingering of strings) are used in performance.

Richard Feynman

Not surprisingly, former Cornell and Caltech Physics Professor and Nobel laureate Richard Feynman played an instrument as well. In this case it was the bongos. In his memoirs Edward Teller describes that while the two were together at Los Alamos, Feynman would play the bongos for hours each evening.[23] Sometimes this would include trips to an area mesa where Feynman was said to have drummed and chanted.[24] Ironically, one of the books about Feynman's life and career is titled, *The Beat of a Different Drum: the Life and Science of Richard Feynman*.[25] Although he was originally self-taught, while at Cornell, Richard Feynman took lessons on the bongos,[26] and immersed himself in Cornell's robust dance community.[27] The Cornell campus and area dance communities remain vibrant to this day.

One of his signature pieces on the bongos was called "Orange Juice," and there are Internet clips (e.g., YouTube) where you can watch Professor Feynman performing in public concerts and small jam sessions. During his time at Caltech, he performed with bongos on stage in the Havana scene of a campus production of the Lerner and Lowe musical, Guys and Dolls[27] and for an internationally-performed ballet written to feature drums.[27] Richard Feynman said that "On the infrequent occasions when I have been called upon in a formal place to play the bongo drums, the introducer never seems to find it necessary to mention that I also do theoretical physics."[28]

Richard Flavell

One of the leading immunology researchers in the world is Richard Flavell of Yale University. He has authored more than 700 papers with most focusing on the area of immune tolerance and autoimmune disease. Thirty years ago Flavell and other Yale-area scientists formed a band called Cellmates.[29] Flavell plays guitar, sings, and writes songs for the band, which has recorded CDs and performs at various events. Additional band members included Ira Melman, the Yale University cell biologist who discovered the endosome. The band's style has been dubbed "Bio-Rock."[29] We were fortunate to have Dr. Flavell accept our invitation to give a recent research seminar at Cornell University. Of course, I should note that Cornell's home (Ithaca, NY) has a quite famous guitar store well worth a guitarist's visit.

NASA Astronauts

A group of NASA astronauts has maintained a long-standing rock band called Max Q that was formed during the dark days of the Challenger

Shuttle disaster.[30] The band rotates performing members based on assignments but has gone through three generations of astronauts and is still going strong. Among the gigs the band has played are performances at the Houston Hard Rock Café and on ABC's TV show *Good Morning America*. The musical repertoire has shifted over the decades to reflect the tastes of band members. Band founder, astronaut Robert Gibson, estimates that 50% of NASA astronauts play musical instruments although a small percentage have joined this specific rock band. Astronaut Jim Weathersby believes that the band has given the astronauts an opportunity to "cultivate their more playful and creative sides."[30]

DANCE

According to Ken Robinson and Lou Aronica in their book, *The Element How Finding Your Passion Changes Everything*,[31] dance, music, science and mathematics are not as different as they may seem. Dance is a highly physical, body-driven process and music is a sound based art form. But as the authors point out, mathematics is something that many dancers and musicians use as an integral part of their performances. Additionally, visual images and pictures are things that scientists and mathematicians often use to test their ideas.[31]

In their article, *Dance Your Experiment*,[32] Michele and Robert Root-Bernstein describe several important experiences and events that emphasize the connection of dance and scientific creativity. Several different types of science-dance connections have been used to interconnect new patterns. In some cases, scientists participated in the dance itself. In other cases, scientists provided the choreography itself drawn from their physics or chemistry or molecular genetic data. Among their stories are the blending of choreography and science.

For example the Root-Bernsteins[32] described how MacArthur Fellow John Cairns, who was working at the Cold Spring Harbor Laboratories, was stuck on a research problem. To address this he had designed a complex experiment that he thought would provide needed information and overcome the present obstacles. Upon describing his experimental plan to a friend who enjoyed folk dancing in the area, the two proceeded to design a "dance of the bacteriophage particles" and then stepped through a series of square dance-type progressive moves along a road between the classroom and lawn. Essentially, they choreographed the experimental design. By converting the experiment to choreography, the two reached the conclusion that the experiment would not work. It would not yield the intended information for Cairns. Cairns accessed a lateral pattern that was a type of kinesthetic analogy for Ness' word analogies.[5] This allowed Cairns to see points connected to the original pattern that, otherwise, had not been available to his perceptions.

For the past few years, one of the most prestigious publications, *Science Magazine*, has sponsored a "Dance Your Ph.D." contest in which videos are submitted for competition. The rules are that each dance has to be based on a scientist's Ph.D. research, and that scientist has to be part of the dance. The annual contest has been highly successful. Students, postdocs and professors can compete against each other. Additionally, there is competition within categories of scientific research (e.g., biology, chemistry, physics). As an example, the 2011 top prize winner for the biology category was Cedric Tan, a biologist at Oxford who presented the dance "Smell Mediated Response to Relatedness of Potential Mates."[33] Of course the overall winner was Joel Miller of the University of Western Australia who presented "Microstructure-Property Relationships in Ti2448 Components Produced by Selective Laser Melting: A Love Story."[33]

In a 2010 article in Science, John Bohannan[34] explored the question "Why Do Scientists Dance" using "Dance Your Ph.D." contestants as a focal point. When Bohannan conducted a survey among participants as to the effect of interpreting their Ph.D.s via dance, he found that there had been mostly positive outcomes if unexpected consequences of interpreting your Ph.D. research in dance form.[34]

Bohannan[34] traced the origins of the science-dance linkage to what was essentially an early type of flash mob dance organized among students by Stanford Nobel Laureate's Paul Berg 1971. This research-based dance included several hundred students, was held on an open field at Stanford, and was called "Dance of the Ribosomes."[35] The students formed a 30S ribosome, brought in messenger RNA, initiation factor 1, transfer RNA, the 50S ribosome and amino acids as they depicted the process of peptide formation and protein synthesis. All of this was set to the rock tune

"Protein Jiva Sutra." A few of the students had some dance experience, knew what was supposed to occur and wore "costumes." However, most of the students were there based on the promise of fun and refreshments. Videos exist of this event and you, too, can select your favorite amino acid from among those performing in this dance.[35]

Dancing and Patterns

I (RRD) have a personal science-dance observation to contribute as well. One of my passions is a dance called West Coast Swing (WCS). It is the state dance of California and considered, along with Argentine tango, to be among the most difficult of dances. Proficient WCS leaders must learn to integrate different musical styles with a myriad of possible dance patterns in a precise manner or risk placing followers in precarious, if not exceptionally unattractive, positions. On the other hand, given the complexity of patterns that exist in the dance, WCS dance followers must be prepared for anything and assume nothing on the dance floor. Once at a national-level WCS dance workshop/competition in Boston, I surveyed a large table of non-professional dancers and asked leaders and followers what they did professionally. Followers showed no pattern at all. Professions ran the gamut. But the leaders were all scientists, mathematicians or engineers. I have since enjoyed conducting this survey at other WCS events with similar outcomes. WCS dancing attracts scientists and may well provide lateral pattern access given the nature of leading this dance.

In a recent article examining dance-driven improvisation titled "Dance as an Algorithm: What Happens When an Animated GIF Comes to Life"[36] the writer/observer details insights gained by using our bodies in dance for new perceptions of technology and Internet interconnections. The article reviews a dance-theater show named "She was a Computer" created by Cara DiFabio that explores technology-human body interfaces using dance. At its heart is one of the themes of this book; the fact that patterns of information connected to you rarely end at your big toe or the top of your head. Instead, they interconnect you to things as mundane as a hair dryer or a touchpad.[36]

In one study, Fink *et al.*[37] examined the level of EEG-based alpha wave sychronization produced during two different dance visualization exercises.

One exercise required individuals to visualize a dance that was completely new and unique (a more creative task) while another asked them to visualize a monotonous pattern of waltz moves. A group of novice dancers was compared again to a group of professional dancers. Both groups were also given the Alternative Uses Test as a measure of creativity. The researchers found that during both the Alternative Uses Test and the improvisational dance visualization exercise, the professional dancers had more right hemispheric alpha synchronization than the novice group. No group differences were observed during the monotonous waltz visualization exercise.[37]

IMPROVISATIONAL THEATER

Improvisational performance of one's science has become a vehicle for both improved science communication and increased opportunities for creative thinking in research. There is evidence suggesting that joint improvisation action, sometimes called "mirror games," can result in a high level of novelty in motion that is also highly synchronized particularly when there is no designated leader.[38] A unique state of togetherness is reached that may be analogous to that attained in some partner dancing.

Recent examples of the application of improvisational theater to science include:

(1) the Alan Alda led, improvisational theater that helps scientists connect with each other and their audiences. This was facilitated by the Center for Communicating Science at SUNY-Stony Brook.
(2) a National Science Foundation (Division of Information and Intelligent Systems) funded pilot project on Improvisational Theater for Computing Scientists.
(3) Science Night at the Improv. Kendall Powell[39] describes the improvisation CellSlam contest that was held at the 2006 meeting of the American Society for Cell Biology (ASCB).

Given the utility of Improv to open up your creative insights, our following exercise provides you with the opportunity to combine Improv and embodied action toward a useful goal. This exercise can be equally effective for individuals as well as groups.

THE MIME IN THE BOX IMPROV EXERCISE

We all have a box. Having a box is not a bad thing by itself. But your box, including its size and characteristics, will affect how you interact with your perceived reality. The question is: is your box the size of a mansion on a cliff overlooking a majestic ocean, more like a comfortable condo-size structure or so small you can barely turn around? Just remember you built your box although you probably had some help from others along the way. It is here that we will give you the chance to redesign your box. A great way to do box renovation is to use improvisational theater. If useful, you can try this in combination with the 30-sec heart meditation discussed earlier in the book.

1. Pretend you are a mime in a box. Feel around and locate the walls, ceiling and floor of the box. Realize that there are no windows or doors in your box.
2. Your first task is to feel around in your box and find something that will allow you to break through the confining walls.
3. Smash the walls, ceiling and floor until the box is no longer there and confining you.
4. Now form another box that better fits the new you and can hold your future aspirations.
5. Make sure you build in doors, windows, skylights, escape hatches — anything that will give you the opportunity to see beyond the four walls and glimpse/access greater possibilities.

HANDCRAFTS

Handcrafts (such as metalwork, woodwork, sculpting, turning pottery items and anything produced by hand in other media) provide an often overlooked avenue for laterally engaging patterns of information. You might encounter these either through handling these items as an admirer or through handiwork production of your own. Of course, in the past handiwork also included writing as this was only done by hand. Using your hands to produce an item or product is one access point to lateral patterns. Was da Vinci a sculptor, a painter or an innovative scientist? Was Goethe a writer, a horticulturalist, an optics scientist or an expert on Renaissance antiquities? The answer probably depends more on your own context for da Vinci and Goethe rather than on their creative capacities or contributions. In the long running U.S. TV show NCIS, the clever lead investigator, Gibbs (played by actor Mark Harmon) seeks the refuge of his basement and his never-ending project of hand hewing wood to build a wooden sailboat. This is the hands-engaged "space" where he puts it all together. Working with a plane in his hand can unlock a plan for solving a crime. Yes, it is fiction but it resonates with viewers because we have a frame-of-reference for occupying our hands and gaining seemingly unconnected insights.

While creating handcrafts can be a toy in our toy box for better access to nonlinear information patterns, similar results can occur by simply handling and enjoying handcrafts. In fact, I (RRD) credit an old Scottish handcrafted silver teapot of particularly unusual size, shape, style, and production. It was an anomaly lined up next to several well-made but standard and infinitely smaller London-made teapots. How could the Scottish teapot be so different? Merely handling that anomalous teapot led to increased curiosity and a detailed study of patterns. In a prior chapter we discussed pattern jumping and how working on patterns in Scottish silver resulted in five books and four research articles concerning Scottish culture and history. That work on non-science patterns around Scotland then led to several papers concerning biomedically-relevant patterns. But the take home message is that the key starting point for engaging these patterns was physically handling that handcrafted 1734–35 Scottish teapot. It not only looked but felt different to me (RRD). If you are drawn to an object, examine it from several vantage points and if permissible, handle it.

Lateral transfer of patterns of information can work that way. One of our toys in the toy box is a modern "Mata Ortiz" vase that has an inspiring multi-DNA strand geometric design. Another looks like DNA restriction fragment banding patterns (Figure 18.1).

The Mata Ortiz pottery is based on ancient Casa Verde (in present-day Mexico) pottery designs. What is the usefulness of the vase? It reminds us to watch for other patterns. Among the other handiwork prompts to alert us to watch for patterns is a set of Matryoshka, nested dolls illustrated in a prior chapter. These dolls have woodwork and painted handcrafting, as well as archetypes, and self-similar fractal dimensions all going for them. Not bad when you want access to patterns.

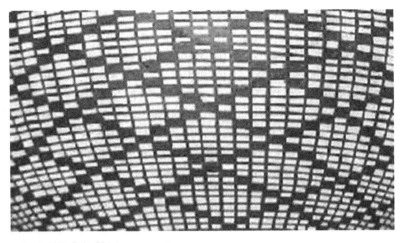

photograph by Janice Dietert

Figure 18.1   Geometric Pattern Similar to DNA Laddering on Pottery from Mata Ortiz

ART AND PHOTOGRAPHY

The connection between art and science is a prime example of the benefits of common tools for: 1) accessing patterns and 2) using lateral patterns to access and organize information. Parallels between art and science have been noted by many authors. In the book, *The Act of Creation*,[40] British author Arthur Koestler presented the case that the capacity for inspiration and thought is enhanced when rationality is suspended. He advocates setting aside automatic behavioral routines to more easily enter dreamlike states where rationality is suspended. Giorgio Careri[41] noted that both physicists and artists have similar processes characterized by "the progressive recognition of a 'sense' in an ensemble of perceived signs."[41] Andre Heck[42] surveyed scientists (largely astronomers) and artists and concluded that there is no unique creativity process. Both groups exhibited major similarities. In the terminology of this book, we might say they both require access to patterns. Similarly, in his book *Hidden Harmony*,[43] J. R. Leibowitz pointed out the themes and understanding that are shared by science and art.

Recently, the Chicago Art Magazine conducted an examination of the intersection between art and science in a series of three articles by Anna

Schier. The first article considered the incorporation of visual art into science-based institutions.[44] Schier believes that coexposure to science and art has a synergistic benefit.[44] Increasingly, art is seen as not only inspirational but also potentially therapeutic. This idea was taken even further by Karlic and Baurek[45] who argued that exposure to art can have epigenetic effects that impact health risks.

This idea of art and well-being has led to its broader incorporation of art in hospitals.[46] Select major research institutions such as the Mayo Clinic have exceptionally rich programs in the visual arts. The Mayo's collection was begun at least as early as 1914 with the completion of its first building.[47] It is so extensive that they have to rotate exhibits much like would occur at a major art museum. Additionally, the Mayo Clinic uses a network of volunteers to provide regular art tours to visitors. For biomedical researchers at the Mayo Clinic, you cannot help but encounter the visual arts on a daily basis.

A second article by Schier[48] considered the reverse relationship, that is the opportunity for museums to function as both science-lessons and art shows. For example, the "Anatomy in the Gallery" section of the International Museum of Surgical Science "consists of pieces that stand as artworks in and of themselves but that also happen to incorporate science, as opposed to serving as art that exists purely to record or demonstrate scientific history."[48] This type of blending of art and science seems likely to invite lateral pattern access.

The final article in the series by Schier is titled "Scientists as Artists."[49] Schier details examples of the Chicago area scientist-artists and the way in which their activities support each other. A consistency is that once artwork has emerged for the scientists, they invariably find ways to introduce their art-based ideas into the research and classroom settings. Schier concludes "The scientist-artist is not a myth, but, rather, a revolutionary creative force. Two methods of thought, which at face value are at odds, can be the catalyst for innovative work."[49]

Do you have to do the active process of painting to increase your access to lateral patterns of information? Not necessarily. Just as when I (RRD) hold a well-crafted teapot, others have already done the painting and you can use that. Observing paintings from a place of nonlinearity and non-rationality can be very useful. In fact observing the following category of art is perhaps the most eye-opening. This category of what has been termed new media art is Fractal Art.

Fractal art truly blends the modalities of mathematics and art. As summarized by Kerry Michell,[50] it is computer generated but under the direction of the artists. The Madelbrot Set that first appeared in 1985 is probably the first example of published fractal art. The fact that it is computer generated should not imply that it is limited and readily predictable to the observer. In fact it is a richly diverse and effective means of accessing unconscious patterns of information. Fractal art and video have been used by psychologists to help patients gain access to unhelpful patterns.[51]

Scott Draves,[52] a fractal artist, described a recent phenomenon in fractal art: the electric sheep domain. Electric Sheep (www.electric sheep.org) is a screen saver program using preferential votes from 50,000 users as a fitness mechanism on an algorithm of abstract animations known as "sheep." The art is all about the "gene pool" for these "sheep." Users can select among computer generated images or incorporate hand-drawn images so crowd-sourcing, or intelligent design, collaborates and competes with evolution among these "sheep." As Draves describes "the images are interference patterns between groups of nonlinear geometric transformations of the plane…. This nonlinear map from genome to image defines a visual language, whose search and incantation brings the machine to life."[52] If you are unsure what electric sheep may do for wool production next winter, there seems to be little question of its effect in facilitating lateral access to patterns.

THE ART IS INFORMATION EXERCISE

Note that I (RRD) first tried this in a perfect science-art juxtaposition. I was at the U.S. National Portrait Gallery within 10 minutes after leaving a panel meeting at the Institute of Medicine, National Academy of Sciences (within view of each other).

1. Walk up to a painting you would normally approach and take a glance at it.
2. Now silently ask the question (Could be directed to yourself, the painting, nothing): Is there any information for me here? (This may sound silly but I assure you it is a painless way to gain insights.)

(Continued)

(*Continued*)

3. Spend at least a minute noticing where your attention goes and what you notice. Don't try to make rational sense of it now. The information may come in any possible form (colors, shapes, words popping into you head). Continue to take note of what images and ideas show up the next few hours. You may want to write them down.

4. Repeat this in front of 5–10 paintings total. (Why do I suggest this range. With fewer you may not get the idea and some paintings may have no information for your access. With more you could have information overload. Of course, I once tested my own limits by looking at about 100 paintings in a row this way and was rewarded with insights plus a headache.)

5. Vary this exercise by instead asking another question while standing in front of paintings: "What would it be like if I knew what is here for me?" This is not a cosmic philosophical question. This is information gathering. You are merely gathering information for your science sifting process that you may previously have ignored. The artists already worked with patterns and those patterns are there in the painting. Go and get them for use with lateral transfer. Again, simply notice either things about the paintings themselves and/or ideas that show up within the next several hours.

MULTIMEDIA ART

Blending science with multiple forms of art can have real benefit. A particularly innovative initiative has been spearheaded by Cornell University Nobel laureate chemist Roald Hoffmann (also author of our Foreword). As a cutting edge chemist, Hoffmann had found diversion and inspiration in poetry and stage work. He authored three series of poems[53–55] as well as coauthored a play titled *Oxygen*.[56] His emerging view of the strong connection between science and art led him to begin a series linking the two more than a decade ago. It is called *Entertaining Science*[57] and is held at the Cornelia Street Café in New York City. The programs pair public lectures by prominent scientists and innovators such as Benoit Mandelbrot and Oliver Sachs with performances by musicians, bands, dancers, poets, artists, actors, photographers who address the lecture's theme so that people can play with ideas, engage science and be entertained. For example the Nov, 4 2007 program was titled "All is Pattern" and paired

NYU environmental scientist, Tyler Volk, with the Christopher Caines Dance Company performing to Eric Satie's music.[57]

For additional exercises that pertain to the use of art to enhance creativity in your workplace, please see the chapter on Creative Spaces.

SUMMARY

Using lateral patterns to access and perceive new scientific linkages is a very powerful tool for innovation in research. While the different media offer a menu of options. Individual researchers should not expect them to have equal personal utility. One person may never get creative glimpses via dance but may discover the greatest scientific insights of a career while playing a musical instrument, viewing or listening to a concert or opera, or standing in front of a painting, photograph or sculpture. Each of these is a gateway. We have only to leave ourselves enough wiggle room to find a few of those *Science Sifting* nuggets of information.

Chapter 19
Summing it Up

How wonderful that we have met with a paradox. Now we have some hope of making progress. — Niels Bohr[1]

We may find illustrations of the highest doctrines of science in games and gymnastics, in travelling by land and by water, in storms of the air and of the sea, and wherever there is matter in motion. — James Clerk Maxwell[2]

Some subjects are so serious that one can only joke about them.
— Niels Bohr[3]

This book was written to emphasize the benefits of tools for scientists and highly-technical professionals beyond those that have been traditionally taught in our university graduate research and postdoctoral training programs. There is a time and place for the highly-focused disciplinary expertise and well-engrained protocols we learned in these programs. But when you hit a roadblock in your research or your overall career needs a boost of vision, the same tools that led you to that brick wall are unlikely to provide you with the solution sets you seek. Yet, that brick wall presents you with a marvelous opportunity to expand your universe of solution sets.

While the tools we have discussed and the exercises we have provided vary significantly from each other, they have a common theme; they help you to shift your perspective by providing you with a different view or vantage point of the problem. Multiple tools are available and multiple vantage points are at your disposal. It is likely that not all the tools will be equally effective for each individual. The satisfaction comes from identifying those tools that work best for you. We encourage you to put them to the test and make them part of your regular routine. Additionally, these tools can be combined for added benefit. For example, meditation and embodied cognition, purposeful language combined with play or sleep techniques linked with synchronicities are examples of combinations you can try.

Finally, we stress the importance of journaling. If you have not had a long history of previously using these tools, you may get resistance from the part of you that has been trained for years to accept only those observations that fit a logical, linear model. You may try to talk yourself out of nonlinear progress. You may ask yourself: Did it really happen? Did it happen that way? In fact, we still have our moments where self-doubt arises. They usually show up when significant progress is at hand much as when moments of confusion precede clarity. If you have recorded what you noticed using these tools, any resistance can be swept aside. You noticed it, you wrote it down, it happened. Your journal will help to feed the more creative you.

Science can be serious but it can also be humorous and fun. It has duality and is more than it might initially seem. Serious tools are really just toys in your game of research. In fact, the lesson from Richard Feynman and others is that maintaining this duality is highly useful. If you look for duality in your work, you will start to open yourself up to seeing your work differently. To help you remember to look for dualities in your work, we will end with a short, slightly-embarrassing personal experience. It reminds us to look further, look beyond, consider things differently, and perceive more broadly.

DO YOU HAVE THAT RECORD?

While in high school in south-central Texas, I (RRD) worked summers at an unairconditioned factory that made cardboard boxes. It was hot, physical work and one where hundreds of daily papers cuts were routine at least for a novice worker like me. At that time, I was very interested in two past-times, dating and music (the latter was reflected by my ever-expanding vinyl record collection). A portion of my income went to support my past-times. In fact as a measure of hours at work, I determined the approximate number of paper cuts it would take for me to have enough disposable income to buy my next recording (the paper cuts to vinyl disc formula). Then came the day when I realized that gaining more enjoyment and insights from my existing records might be more helpful than mindlessly buying yet another record just because I could. Don't be surprised if this reminds you of some classic comedy routine.

The place to go in San Antonio, TX in the 1960s for all things musical (other than sheet music) was San Antonio Music. It was located in downtown and the store handled instruments and instrument repairs, music stands, band accessories, and an impressive collection of vinyl records of all music types. This was pre-desktop computers and personal Internet access so in-store stock was a prime consideration. If you wanted a top-40s pop song or an obscure opera recording, this was the best area source. Since it was a long trek to get to the store from my home, I wanted to phone ahead and see if they had the new record I wanted among their rows and rows of recordings. The phone conversation went as follows:

Call is answered: rapid, muffled greeting blurted out before the receiver is in place.

Me: Can you please connect me to your record department?

Them: Yes, a moment please. (appropriate time lag for extension connection). 'record department'

Me: Can you tell me if you have in stock the recording of ....?"

Them: This is the 'record department.'

Me: I know. I am looking for the recording of ......"

Them: But we are the 'record department.'

Me: Precisely. This recording was produced in ... and is .....

Them: I don't think we carry anything like that.

Me: Well if helpful, I think the last time I visited I noticed the appropriate section is over in the far right corner of your first floor.

Them: That is where we have our 'Fenders.'

Me: Yes but I don't play the guitar. I am looking for the (year) recording of.....

Them: (after a long, uncomfortable pause). This is the Wrecker Department of (car dealership).

What a surreal moment. I learned several lessons from this experience: (1) I did not really need a new recording despite my impressive number of paper cuts for the week. Instead, it was more useful to re-explore and delve deeper into the recordings I already had. (2) Synchronicities happen. (3) The Universe really is a funny place. (4) There is always more than one possibility. My Abbot and Costello moment is useful whenever I question whether the tools, toys and information I need are within my grasp.

SUMMARY

More important than gathering more of the same information or acquiring more of anything is the ability to discern the usefulness of data and objects in the moment. Remaining in one, highly focused vantage point will not always allow you to gain the scope of view necessary to discern the data or its usefulness or applicability to the problem. Having multiple vantage points and various ways of gathering data gives you the broader perspective necessary for well-developed discernment. Discernment, in turn, will help open roadblocks, give you greater perspective on your work and overall help facilitate greater creative innovations. Multiple vantage points are more likely to bring you to that moment of inspired discovery than will a straight, focused, doggedly-determined path of ever-narrowing research.

References

Introduction

1. Holmes, O. W., *The Professor at the Breakfast Table.* In Project Guttenberg EBook #2665 ed.; 1860; X The Book of the Three Maiden Sisters. Paragraph 9 line 2, http://www.gutenberg.org/files/2665/2665-h/2665-h.htm#link2H_ PREF. Accessed November 7, 2012.
2. Wolff, G. *The Next Insanely Great Thing.* The Wired Interview Periodical [Online], 1996, p. 8. http://www.wired.com/wired/archive/4.02/jobs.html. Accessed November 7, 2012.
3. Wallis, C.; Ferenbaugh, D.; Johnson, M., *Honoring a Modern Mendel.* Time Magazine 1983, page 43.
4. von Oech, R., *A Whack on the Side of the Head.* Business Plus: New York, 2008; 256pp.

Chapter 1

1. Medawar, P. B., *Induction and Intuition in Scientific Thought. Jane Lectures for 1968.* American Philosophical Society: Philadelphia, 1969; 62pp page 59.
2. Faraday, M., *Experimental Researches in Chemistry and Physics.* Richard Taylor and William Francis: London, 1859; 496pp page 469.
3. Rosanoff, M. A., *Edison in His Laboratory.* Harper's Magazine September, 1932, page 403.
4. Root-Bernstein, R.; Root-Bernstein, M., *Sparks of Genius.* Mariner Books: Boston, 2001; 416pp.
5. Bartlett, R., *The Physics of Miracles.* Atria Books: New York, 2009; 288pp.
6. Ness, R. B., *Innovation Generation.* Oxford University Press: Oxford, 2012; 272pp.

7. Gladwell, M., *Blink: The Power of Thinking Without Thinking*. Back Bay Book: New York, 2007; 296pp.
8. von Oech, R., *A Whack on the Side of the Head*. Business Plus: New York, 2008; 256pp.

Chapter 2

1. Kostelanetz, R., *Conversing With Cage*. 2nd ed.; Routledge: New York, 2003; 344pp page 221.
2. Gladwell, M., *Blink: The Power of Thinking Without Thinking*. Back Bay Book: New York, 2007; 296pp.
3. Maurois, A., *The Life of Sir Alexander Fleming: Discoverer of Penicillin*. Pengiun Books: New York, 1963; 328pp page 120.
4. Root-Bernstein, R.; Root-Bernstein, M., *Sparks of Genius*. Mariner Books: Boston, 2001; 416pp pages 246, 248.
5. von Oech, R., *A Whack on the Side of the Head*. Business Plus: New York, 2008; 256pp.
6. Feynman, R. P.; Leighton, R.; Hutchings, E., *Surely You're Joking, Mr. Feynman!* W.W. Norton & Company: New York, 1985; 350pp pages 137–155.
7. Feynman, R. P.; Leighton, R.; Hutchings, E., *Surely You're Joking, Mr. Feyman!* W.W. Norton & Company: New York, 1985; 350pp page 173.
8. Lang, S., *CU plate in Nobel Prize exhibits symbolizes role of playfulness in creativity*. Cornell Chronicle March 24, 2005, page 5, http://www.news.cornell.edu/chronicle/05/3.24.05/CUplate.html. Accessed November 5, 2012.
9. Lovell, B., *The Cavity Magnetron in World War II: Was the Secrecy Justified?* Rec. R. Soc. Lond. 2004, 58, 283–294 http://rsnr.royalsocietypublishing.org/content/58/3/283.full.pdf
10. Earls, A. R.; Edwards, R. E., *Raytheon Company: The First Sixty Years*. Arcadia Publishing, 2005; 128pp.
11. Ratheon. *History of the Ratheon Corporation*. http://www.raytheon.com/ourcompany/history/leadership/. Accessed November 11, 2012.
12. Acton, J.; Adams, T.; Packer, M., *Origin of Everyday Things*. Sterling: New York, 2006; 320pp page 156.
13. Keller, E. F., *A Feeling for the Organism: The Life and Work of Barbara McClintock*. 10th Anniversary ed.; Times Books: New York, 1984; 272pp page 70.

14. Kass, L. B.; Chomet, P., *Barbara McClintock*. In Maize Handbook — Volume II: Genetics and Genomics, Benntzen, L.; Hake, S. C., Eds. Springer: New York, 2009; 17–52pp, page 28.

15. Keller, E. F., *A Feeling for the Organism: The Life and Work of Barbara McClintock*. 1984; page 142.

16. Keller, E. F., *A Feeling for the Organism: The Life and Work of Barbara McClintock*. 10th Anniversary ed.; Times Books: New York, 1984; page 198.

17. Keller, E. F., *A Feeling for the Organism: The Life and Work of Barbara McClintock*. 10th Anniversary ed.; 1984 Times Books: New York, page 36.

18. Comfort, N. C., *The Tangled Field: Barbara McClintock's Search for the Patterns of Genetic Control* Harvard University Press: Camridge, MA, 2003; 368pp.

19. Lian, C., *Rediscovering Barbara McClintock*. The Cornell Daily Sun September 24, 2008, 2008, http://cornellsun.com/node/32029. Accessed October 20, 2012.

20. Lakin, P., *Steve Jobs: Thinking Differently*. Simon & Schuster: New York, 2011; 192pp.

21. Doeden, M., *Steve Jobs: Technology Innovator and Apple Genius*. Lerner Publishing Group: Minneapolis, MN, 2012; 48pp.

22. Gilliam, S., *Steve Jobs: Apple & Ipod Wizard*. ABDO Publishing Company: Edina, MN, 2008; 112pp.

23. Elliot, J.; Simon, W. L., *The Steve Jobs Way: iLeadership for a New Generation*. Vanguard Press: New York, 2011; 256pp.

24. Stevens, C., *Designing for the iPad: Building Applications that* Sell. Wiley: Chichester, UK, 2011; 352pp, page 34.

25. amazon.com, *An Interview with Walter Isaacson.* http://www.amazon.com/Steve-Jpbs-Walter-Isaacson/dp/1451648537. Accessed October 12, 2012.

Chapter 3

1. Einstein, A.; Calaprice, A.; Dyson, F., *The Ultimate Quotable Einstein*. Princeton University Press: Princeton, NJ, 2011; 576pp page 386.

2. De Bono, E., *I am Right You are Wrong: From This to the New Renaissance: From Rock Logic to Water Logi*. Viking/Penguin: New York, 1991; 320pp page 17.

3. Dobzhansky, T. G., *The Biological Basis of Human Freedom*. Columbia University Press: New York, 1956; 139pp.

4. Jobs, S., *Commencement Address at Stanford University*. In Palo Alto, CA, given June 12, 2005; http://news.stanford.edu/news/2005/june15/jobs-061505. html. Accessed November 6, 2012.

5. Project, T. O. C. M. *Directory Index: Edsel/1958_Edsel/1958_Edsel_Sell-O-Graph*. http://www.oldcarbrochures.com/static/NA/Edsel/1958_Edsel/1958_ Edsel_Sell-O-Graph/1958%20Edsel%20Sell-O-Graph-04a.html. Accessed July 5, 2012.

6. Caprice, A.; Dyson, F.; Albert, E., *The New Quotable Einstein*. Princeton University Press: Princeton, NJ, 2005; 440pp page 69.

7. Ragins, B. R.; Kram, K. K., *The Handbook of Mentoring at Work: Theory, Research, and Practice*. Sage Publications: Thousand Oaks, CA, 2007; 760pp.

8. Allen, T. D.; Eby, L. T., *The Blackwell Handbook of Mentoring: A Multiple Perspectives Approach*. Wiley-Blackwell: Malden, MA, 2007; 520pp.

9. Beck, M., *Steering by Starlight*. Rodale Books: Emmaus, PA, 2008; 256pp.

10. Forrest, H.; Sutton, H. E.; Riggs, A., *In Memorium: Burke Hancock Judd*. In 2007–2008 Memorials, Council, O. o. t. G. F. F., Ed. The University of Texas at Austin: Austin, TX, 2008; http://www.utexas.edu/faculty/council/2007-2008/ memorials/judd.html, August 4, 2012.

Chapter 4

1. Palahniuk, C., *Survivor: A Novel*. W.H. Norton & Company: New York, 2010; 289pp page 118.

2. Sheff, D., *Playboy Interview: Steven Jobs*. Playboy Magazine February, 1985, http://www.scribd.com/doc/43945579/Playboy-Interview-with-Steve-Jobs. Accessed October 24, 2012.

3. Reading, A., *Meaningful Information: The Bridge Between Biology, Brain, and Behavior*. Springer: New York, 2011; 172pp pages 1–2.

4. Henriques, G., *A New Unified Theory of Psychology*. Springer: New York, 2011; 307pp page 70.

5. Riggs, A., *Deep Tissue Massage, Revised: A Visual Guide to Techniques*. North Atlantic Books: Berkely, CA, 2007; 307pp, page 107.

6. Upledger, J., *Releasing Emotions Trapped in the Tissues*. Massage Today 2008, http://massagetoday.com/mpacms/mt/article.php?id=13825. Accessed July 16, 2012.

7. Mandelbrot, B. B., *The Fractal Geometry of Nature*. W. H. Freeman & Company: New York, 1982; 480pp.

8. Esteban, F. J.; Padilla, N.; Sanz-Cortes, M.; de Miras, J. R.; Bargallo, N.; Villoslada, P.; Gratacos, E., Fractal-dimension analysis detects cerebral changes in preterm infants with and without intrauterine growth restriction. *Neuroimage* 2010, 53, (4), 1225–32.

9. Cheung, N.; Liew, G.; Lindley, R. I.; Liu, E. Y.; Wang, J. J.; Hand, P.; Baker, M.; Mitchell, P.; Wong, T. Y., Retinal fractals and acute lacunar stroke. *Ann Neurol* 2010, 68, (1), 107–11.

10. Ionescu, C. M.; Muntean, I.; Tenreiro-Machado, J. A.; De Keyser, R.; Abrudean, M., A theoretical study on modeling the respiratory tract with ladder networks by means of intrinsic fractal geometry. *IEEE Trans Biomed Eng* 2010, 57, (2), 246–53.

11. Dietert, R. R., Fractal immunology and immune patterning: potential tools for immune protection and optimization. *J Immunotoxicol* 2011, 8, (2), 101–10.

12. Wu, X.; Huan, T.; Pandey, R.; Zhou, T.; Chen, J. Y., Finding fractal patterns in molecular interaction networks: a case study in Alzheimer's disease. *Int J Comput Biol Drug Des* 2009, 2, (4), 340–52.

13. West, G. B.; Enquist, B. J.; Brown, J. H., A general quantitative theory of forest structure and dynamics. *Proc Natl Acad Sci USA* 2009, 106, (17), 7040–5.

14. Zipf, G., *Human Behavior and the Principle of Least Effort: An Introduction to Human Ecology.* Facsimile Edition ed.; Martino Fine Books: 2012; 588pp.

15. Glaeser, E. L., *A Tale of Many Cities.* Periodical [Online], 2010. http://economix.blogs.nytimes.com/2010/04/20/a-tale-of-many-cities/. Accessed November 11, 2012.

16. Easley, D.; Kleingberg, J., *Networks, Crowds, and Markets: Reasoning About a Highly Connected World.* Cambridge University Press: Cambridge, 2010; 744pp.

Chapter 5

1. Curie, E., *Madame Curie : A Biography.* Reissue ed.; Da Capo Press: Cambridge, MA, 2001; 448pp page 341.

2. Dietert, R. R., *Notes on "Creativity Examined" Workshop.* At Cornell University: Ithaca, NY, August 1, 2012.

3. Vaughn, R., *Listen to the Music: The Life of Hilary Koprowski.* Springer: New York, 2000; 320pp pages 83–84.

4. Pyke, D., *Pyke's Notes*. Royal College of Physicians of London: London, 1992; 295pp page 167.

5. Medawar, P. B., *The Uniqueness of the Individual*. 2nd revised edition ed.; Peter Smith Pub Inc: Gloucester MA, 1987.

6. Medawar, P. B.; Medawar, J. S., *Aristotle to Zoos. A Philisophical Dictionary of Biology*. Harvard University Press: Cambridge, MA, 1985; 320pp.

7. Medawar, P. B., *Memoir of a Thinking Radish: An Autobiography*. Oxford University Press: Oxford, 1988; 224pp.

8. Lee, R., *The Eureka! Moment: 100 Key Scientific Discoveries of the 20th Century*. Routledge: New York, 2002; 256pp pages 32–33.

9. von Oech, R., *A Whack on the Side of the Head*. Business Plus: New York, 2008; 256pp.

10. Berkun, S., *The Myths of Innovation*. O'Reily Media: Sebastopol, CA 2010; 248pp.

11. Ness, R. B., *Innovation Generation*. Oxford University Press: Oxford, 2012; 272pp.

12. Root-Bernstein, R.; Root-Bernstein, M., *Sparks of Genius*. Mariner Books: Boston, 2001; 416pp pages 246, 248.

Chapter 6

1. Fitzgerald, F. S., *The Great Gatsby*. Reissue ed.; Scribner: New York, 2004; page 132.

2. Meyer, J., *Change Your Words, Change Your Life: Understanding the Power of Every Word You Speak*. FaithWords: New York, 2012; 320pp.

3. Ness, R. B., *Innovation Generation*. Oxford University Press: Oxford, 2012; 272pp.

4. Bartlett, R., *The Physics of Miracles*. Atria Books: New York, 2009; 288pp.

5. Harper, D., *Novel*. http://www.etymonline.com/index.php?term=novel. Accessed October 24, 2012.

6. Carter, R., *Language and Creativity: The Art of Common Talk*. Routledge: London, 2004; 272pp.

7. Beck, M., *The Four Day Win*. Rodale Press: Emmaus, PA, 2008; 384pp, pages 120–122.

8. Dietert, J., *Notes on the Lecture of Dr. Richard Bartlett*. In Denver, CO, August 2008.

9. School, R. H. L. *English Basics* http://www.rhlschool.com/eng3n26.htm. Accessed November 20, 2012.

10. Grothe, M., I *Never Metaphor I Didn't Like: A Comprehensive Compilation of History's Greatest Analogies, Metaphors, and Simile.* HarperCollins: New York, 2008; 366pp.

11. Anonymous. *Metaphors: Dr. Mardy Grothe Knows Them All.* Periodical [Online], 2008. http://www.thepilot.com/news/2008/nov/30/metaphors. Accessed November 11, 2012.

12. Lakoff, G.; Johnson, M., *Metaphors We Live By.* University of Chicago Press: Chicago, 1980; pages 193, 235.

13. Norquist, R. *Using Similes and Metaphors to Enrich Our Writing. Dead Metaphors.* Periodical [Online], 2012. http://grammar.about.com/od/d/g/deadmetterm.htm. Accessed September 26, 2012.

14. Lakoff, G.; Johnson, M., *Metaphors We Live By.* University of Chicago Press: Chicago,1980; pages 7–8.

15. Hall, D., *Innovation in English Language Teaching: A Reader.* Routledge: New York, 2001; 304pp.

16. Pascolini, A.; Pietroni, M., Feynman diagrams as metaphors: borrowing the particle physicist's imagery for science communication purposes. *Physics Education* 2002, 37, 324.

17. Sagarin, S. K.; Gruber, H. G., *Ensemble of Metaphor.* In Encyclopedia of Creativity Runco, M. A.; Pritzker, S. R., eds. Academic Press: San Diego, 1999; Vol. 1, pages 677–680.

Chapter 7

1. Adams, D., *Hitchiker's Guide to the Galaxy: Further Radio Scripts.* Pan MacMillan: 2012; 368pp page 292.

2. Einstein, A., *Dated Letter to the Besso Family.* In Einstein Archives Online: 1955; http://alberteinstein.info/vufind1/Record/EAR000050096. Accessed November 1, 2012.

3. da Vinci, L.; Richter, I. A.; Kemp, M., *Notebooks.* Oxford University Press: Oxford, 2008; 352pp page 6.

4. Poe, E. A., *A Dream within a Dream.* Flag of Our Union volume 4 section 2, 1849, page 13.

5. Schrodinger, E., *What is Life?: With Mind and Matter and Autobiographical Sketches*. Cambridge University Press: Cambridge, 2012; 194pp page 122.

6. Purves, D.; Wojtach, W. T.; Lotto, R. B., Understanding vision in wholly empirical terms. *Proc Natl Acad Sci USA* 2011, 108 Suppl 3, 15588–95 see also Meredith, D., *Tricking the Eye or Tapping a Reflex?: Revisiting Vision*, Duke Magazine, 2000, July–August issue, pages 14–19.

7. Bartlett, R., *The Physics of Miracles*. Atria Books: New York, 2009; 288pp page 164.

8. Hunter, J. P.; Katz, J.; Davis, K. D., Stability of phantom limb phenomena after upper limb amputation: a longitudinal study. *Neuroscience* 2008, 156, (4), 939–49.

9. Martin Ginis, K. A.; McEwan, D.; Josse, A. R.; Phillips, S. M., Body image change in obese and overweight women enrolled in a weight-loss intervention: the importance of perceived versus actual physical changes. *Body Image* 2012, 9, (3), 311–7.

10. Chabris, C.; Simons, D., *The Invisible Gorilla: How Our Intuitions Deceive Us*. Crown Publishing: New York, 2009; 320pp.

11. Chabris, C.; Simons, D. *the invisible gorilla media web page*. http://www.theinvisiblegorilla.com/gorilla_experiment.html. Accessed November 12, 2012.

12. Boehnke, S. E.; Berg, D. J.; Marino, R. A.; Baldi, P. F.; Itti, L.; Munoz, D. P., Visual adaptation and novelty responses in the superior colliculus. *Eur J Neurosci* 2011, 34, (5), 766–79.

13. Ness, R. B., *Innovation Generation*. Oxford University Press: Oxford, 2012; 272pp.

14. Duncan, K.; Sadanand, A.; Davachi, L., Memory's penumbra: episodic memory decisions induce lingering mnemonic biases. *Science* 2012, 337, (6093), 485–7.

Chapter 8

1. Feynman, R. P.; Leighton, R.; Hutchings, E., *Surely You're Joking, Mr. Feynman!* W. W. Norton and Company: New York, 1985; 350pp page 36.

2. Nietzsche, F., *Thus Spoke Zarathustra*. Chapter IV. The Despisers of the Body. Oxford University Press: Oxford, 2005; 384pp page 30.

3. Leider, E. W., *Becoming Mae West*. Da Capo Press: Cambridge, MA, 2000; 480pp page 39.

4. Wilson, R. A.; Foglia, L., *Embodied Cognition*. In The Stanford Encyclopedia of Philosophy Fall 2011 ed.; Zalta, E. N., ed. Center for Study of Language and Information, Stanford University: Stanford, CA, 2011, http://plato.stanford. edu/archives/fall2011/entries/embodied-cognition/. Accessed August 9, 2012.

5. Barsalou, L. W., Grounded cognition. *Annu Rev Psychol* 2008, 59, 617–45.

6. Rupert, R. D., *Cognitive Systems and the Extended Mind*. Oxford University Press: Oxford, 2009; 288pp.

7. Isanski, B.; West, C. *The Body of Knowledge: Understanding Embodied Cognition*. Periodical [Online], 2010. http://www.psychologicalscience.org/ observer/getArticle.cfm?id=2606. Accessed August 8, 2012.

8. Zhong, C. B.; Leonardelli, G. J., Cold and lonely: does social exclusion literally feel cold? *Psychol Sci* 2008, 19, (9), 838–42.

9. Williams, L. E.; Bargh, J. A., Experiencing physical warmth promotes interpersonal warmth. *Science* 2008, 322, (5901), 606–7.

10. Kang, Y.; Williams, L. E.; Clark, M. S.; Gray, J. R.; Bargh, J. A., Physical temperature effects on trust behavior: the role of insula. *Soc Cogn Affect Neurosci* 2011, 6, (4), 507–15.

11. Bargh, J. A.; Shalev, I., The substitutability of physical and social warmth in daily life. *Emotion* 2012, 12, (1), 154–62.

12. Miles, L. K.; Karpinska, K.; Lumsden, J.; Macrae, C. N., The meandering mind: vection and mental time travel. *PLoS One* 2010, 5, (5), e10825.

13. Miles, L. K.; Nind, L. K.; Macrae, C. N., Moving through time. *Psychol Sci* 2010, 21, (2), pages 222–223.

14. Clark, A.; Chalmers, D., The Extended Mind. *Analysis* 1998, 58, (1), 7–19.

15. Menary, R., *The Extended Mind*. MIT Press: Cambridge, MA, 2010; 424pp.

16. Dimotakis, N.; Conlon, D. E.; Ilies, R., The mind and heart (literally) of the negotiator: personality and contextual determinants of experiential reactions and economic outcomes in negotiation. *J Appl Psychol* 2012, 97, (1), 183–93.

17. Fink, R. A., *Creative Imagery: Discoveries and Inventions in Visualization*. Taylor & Francis: New York, 1990; 192pp.

18. Morris, T.; Spittle, M.; P., W. A., *Imagery In Sport*. Human Kinetics: Champaign, IL, 2005; 387pp.

19. Clark, L. V., Effect of mental practice on the development of a certain motor skill. *Research Quarterly* 1960, 31, (4), 560–569.

20. Ranganathan, V. K.; Siemionow, V.; Liu, J. Z.; Sahgal, V.; Yue, G. H., From mental power to muscle power–gaining strength by using the mind. *Neuropsychologia* 2004, 42, (7), 944–56.

21. Reiser, M.; Busch, D.; Munzert, J., Strength gains by motor imagery with different ratios of physical to mental practice. *Front Psychol* 2011, 2, 194.

22. Eerland, A.; Guadalupe, T. M.; Franken, I. H.; Zwaan, R. A., Posture as index for approach-avoidance behavior. *PLoS One* 2012, 7, (2), e31291.

23. Hartmann, M.; Farkas, R.; Mast, F. W., Self-motion perception influences number processing: evidence from a parity task. *Cogn Process* 2012, 13 Suppl 1, S189–92.

24. Mertz, A., *The Body Can Speak: Essays on Creative Movement Education with Emphasis on Dance and Drama.* Southern Illinois University Press: Carbondale, IL, 2002; 168pp.

25. Leung, A. K.; Kim, S.; Polman, E.; Ong, L. S.; Qiu, L.; Goncalo, J. A.; Sanchez-Burks, J., Embodied metaphors and creative "acts". *Psychol Sci* 2012, 23, (5), 502–9.

26. Barron, C.; Alton, B., *The Creativity Cure: A Do-It-Yourself Prescription for Happiness.* Simon & Schuster: New York, 2012; 320pp pages 132–133.

27. Mondloch, C. J., Sad or fearful? The influence of body posture on adults' and children's perception of facial displays of emotion. *J Exp Child Psychol* 2012, 111, (2), 180–96.

28. Kessler, K.; Miellet, S., Perceiving Conspecifics as Integrated Body-Gestalts Is an Embodied Process. *J Exp Psychol Gen* 2012, doi:10.1037/a0029617.

29. Eerland, A.; Guadalupe, T. M.; Zwaan, R. A., Leaning to the left makes the Eiffel Tower seem smaller: posture-modulated estimation. *Psychol Sci* 2011, 22, (12), 1511–4.

30. Dijkstra, K.; Kaschak, M. P.; Zwaan, R. A., Body posture facilitates retrieval of autobiographical memories. *Cognition* 2007, 102, (1), 139–49.

31. Scheumann, D. W., *The Balanced Body: A Guide to Deep Tissue and Neuromuscular Therapy.* Lippincott Williams & Wilkins: Baltimore, MD, 2006; 272pp.

32. Shiota, M. N.; Kalat, J. W., *Emotion.* Wadsworth Centage Learning: Belmont, CA, 2011; 454pp.

33. Zhang, W. J.; Yang, X. B.; Zhong, B. L., Combination of acupuncture and fluoxetine for depression: a randomized, double-blind, sham-controlled trial. *J Altern Complement Med* 2009, 15, (8), 837–44.

34. St. George, V.; Lenarz, M., *The Chiropractic Way: How Chiropractic Care Can Stop Your Pain and Help You Regain Your Health Without Drugs or Surgery.* Random House Digital, Inc: New York, 2003; 384pp.

35. Sills, F., *Foundations in Craniosacral Biodynamics,* Volume One. North Atlantic Books: Berkely, CA, 2012; 424pp.

36. Riggs, A., *Deep Tissue Massage, Revised: A Visual Guide to Techniques.* North Atlantic Books: Berkely, CA, 2007; 307pp page 107.

37. Lynch, E., Emotional acupuncture. *Nurs Stand* 2007, 21, (50), 24–5.

38. Dixon, M. W., *Myofacial Massage.* Lippincott Williams & Wilkins: Baltimore, MD, 2006; 240pp page 194.

39. Fernandez-Perez, A. M.; Peralta-Ramirez, M. I.; Pilat, A.; Villaverde, C., Effects of myofascial induction techniques on physiologic and psychologic parameters: a randomized controlled trial. *J Altern Complement Med* 2008, 14, (7), 807–11.

40. Monti, D. A.; Stoner, M. E.; Zivin, G.; Schlesinger, M., Short term correlates of the Neuro Emotional Technique for cancer-related traumatic stress symptoms: a pilot case series. *J Cancer Surviv* 2007, 1, (2), 161–6.

41. Jensen, A. M., A mind-body approach for precompetitive anxiety in power-lifters: 2 case studies. *J Chiropr Med* 2010, 9, (4), 184–92.

42. Upledger, J., *Releasing Emotions Trapped in the Tissues.* Massage Today 2008, http://massagetoday.com/mpacms/mt/article.php?id=13825. Accessed July 16, 2012.

43. Upledger, J. E., *SomatoEmotional Release: Deciphering the Language of Life.* North Atlantic Books: Berkeley, CA, 2002; 312pp.

Chapter 9

1. Viereck, G. S., *What Life Means to Einstein. An Interview with George Sylvester.* Saturday Evening Post October 26, 1929, see page 17.

2. Alda, A., *Things I Overheard While Talking to Myself.* Random House Digital, Inc.: New York, 2008; 256pp pages 21–22.

3. Nadler, G.; Chandon, W., *Smart Questions: Learn to Ask the Right Questions for Powerful Results.* John Wiley & Sons: San Francisco, 2004; 320pp page 99.

4. Ness, R. B., *Innovation Generation.* Oxford University Press: Oxford, 2012; 272pp.

5. IPCS, *Harmonization Project Document No. 10. Guidance for Immunotoxicity Risk Assessment for Chemicals*; World Health Organization: Geneva, 2012; Research, http://www.inchem.org/documents/harmproj/harmproj/harmproj10. pdf. Accessed November 12, 2012.

6. Horrobin, D. F., The philosophical basis of peer review and the suppression of innovation. *JAMA* 1990, 263, (10), 1438–41.

7. Hume, D., *Political Discourses*. R. Fleming printer ed.; A. Kincaid and A. Donaldson: Edinburgh, 1752.

8. Brown, K.; Dietert, R., The Act of Union, the Forty-five, and Scottish Domestic Silverware. *History Scotland Magazine* 2010, 10; (5), pages 27–37.

9. Gladwell, M., *Outliers: The Story of Success*. Little Brown and Company: New York, 2008; 320pp.

10. Brown, K.; Nasilowski, D., Of Robert Barker, Calton Hill and the Scottish Origins of Visual Mass Media. *History Scotland Magazine* 2010, 10; (6) pages 30–35.

11. Gladwell, M., *Blink: The Power of Thinking Without Thinking*. Back Bay Book: New York, 2007; 296pp pages 197–206.

12. Ekman, P.; Friesen, W. V., *Facial Action Coding System: Investigator's Guide*. Consulting Psychologists Press: Palo Alto, CA, 1978; 153pp.

Chapter 10

1. Carlin, G., *Brain Droppings*. Hyperion: New York, 1998; 272pp page 72.

2. Eitrem, D. *Photo Tips — To Get Creative With Your Photography, Shoot Your Photos From A New Vantage Point!* Periodical [Online], 2012. ezine. http://EzineArticles.com/7126440. Accessed August 10, 2012.

3. Belt, A. F., *The Elements of Photography: Understanding and Creating Sophisticated Images*. 2nd ed.; Focal Press, Taylor & Francis: Waltham, MA, 2011; 404pp page 58.

4. Fichner-Rathus, L., *Foundations of Art and Design*. Wadsworth Cengage Learning: Boston, MA, 2011; 360pp page 162.

Chapter 11

1. Moore, R. D.; Braga, B., *All Good Things*. In Star Trek: The Next Generation, Paramount Studios: 1994; Season 7 #40277–747.

2. Zeilic, M., *Concept Mapping in: Field-Testing Learning Assessment Guide.* In Field-Tested Learning Assessment Guide (FLAG) University of New Mexico: 2011; http://www.flaguide.org/cat/conmap/conmap1.php. Accessed December 4, 2011.

3. Novak, J. D., *A Theory of Education.* Cornell University Press: Ithaca, NY, 1977; 295pp.

4. Novak, J. D.; Musonda, D., A Twelve-Year Longitudinal Study of Science Concept Learning. *Amer Educ Res J* 1991, 28, (1), 117–153.

5. Ausubel, D. P., *The Psychology of Meaningful Verbal Learning.* Grune and Stratton: New York, 1963; 272pp.

6. Ausubel, D. P., *Educational Psychology: A Cognitive View.* Holt, Rinehart and Wilson: New York, 1968; 733pp.

7. Novak, J. D., *Learning, Creating, and Using Knowledge: Concept Maps as Facilitative Tools in Schools and Corporations.* Lawrence Erlbaum Associates: Mahway, NJ, 1998; 251 pp.

8. Cañas, A. J.; Ford, K. M.; Novak, J. D.; Hayes, P.; Reichherzer, T.; Suri, N., Online Concept Maps: Enhancing Collaborative Learning by Using Technology With Concept Maps. *The Science Teacher* 2001, 68, (4), 49–51.

9. Novak, J. D.; Canos, A. J., *The Theory Underlying Concept Maps and How to Construct and Use Them;* Florida Institute for Human and Machine Cognition: Pensacola, FL, 2008; Research, http://cmap.ihmc.us/publications/researchpapers/theorycmaps/theoryunderlyingconceptmaps.htm. Accessed November 4, 2012.

10. Edmondson, K. M.; Smith, D. F., Concept mapping to facilitate veterinary students' understanding of fluid and electrolyte disorders. *Teach Learn Med* 1998, 10, (2), 21–33.

11. Dietert, R. R.; DeWitt, J. C.; Germolec, D. R.; Zelikoff, J. T., Breaking patterns of environmentally influenced disease for health risk reduction: immune perspectives. *Environ Health Perspect* 2010, 118, (8), 1091–9.

12. Dietert, R. T.; Zelikoff, J. T., Pediatric immune dysfunction and health risks following early-life immune insult. *Curr Pediatr Res* 2009, 5, 35–51.

13. Dietert, R. R.; Dewitt, J. C.; Luebke, R. W., *Reducing the prevalence of chronic diseases.* In Immunotoxity, Immune Dysfunction and Chronic Disease, Dietert, R. R.; Luebke, R. W., Eds. Springer: New York, 2012; pages 419–440.

14. Wheeldon, J.; J, F., Framing experience: concept maps, mind maps, and data collection in quantitative research. *International Journal of Qualitative Methods* 2009, 8, (3), 68–83.

15. Nesbit, J. C.; Adesope, O.O., Learning with concept and knowledge maps, a meta-analysis. *Rev Educ Res* 2006, 76, (3), 413–448.

16. Iuli, R. J.; Hellden, G., *Using concept maps as a research tool in science education research.* In Concept Maps: Theory, Methodology, Technology, Canas, A.; Novak, J. D.; Gonzalez, F. M., Eds. Proc. Of the First Int. Conference on Concept Mapping: Pamplona Spain, 2004.

17. Iuli, R. J., *The use of metacognition tools in a multidimensional research project.* Cornell University, Ithaca, NY, 1998.

18. Legard, R.; J., K.; K, W., *In-depth interviews.* In Qualitative Research Practice: A Guide for Social Research Students and Researchers Ritchie, J.; Lewis, J., Eds. Sage: Thousand Oaks, CA, 2003; pages 138–169.

19. Wheeldon, J., *Is a Picture Worth a Thousand Words? Using Mind Maps to Facilitate Participant Recall in Qualitative Research.* The Qualitative Report 2011, 16, (2), 509–522 http://www.nova.edu/ssss/QR/QR16-2/wheeldon.pdf.

20. Dietert, R. R.; M., D. J., *Compendium of Scottish Silver.* Internet First University Press: Ithaca, NY, 2006; http://ecommons.library.cornell.edu/bitstream/1813/3026/4/Silver%20Rev%20vol%20one%20smaller%20R6-06.pdf http://ecommons.library.cornell.edu/bitstream/1813/3027/6/Silver%20Rev%20vol%20two%20smaller%20REV6-063.pdf,

21. Dietert, R.; Dietert, J., *Compendium of Scottish Silver II.* Dietert Publications: Lansing, NY, 2007; 632pp.

22. Dietert, R. R.; J.M., D., *The Edinburgh Goldsmiths I. Training, Marks, Output and Demographics.* Dietert Publications: Lansing, NY, 2007; 200pp.

23. Dietert, R.; Dietert, J., *Scotland's Families and the Edinburgh Goldsmiths.* Dietert Publications: Lansing, NY, 2008; 304pp.

24. Brown, K.; Dietert, R., The Act of Union, the Forty-five, and Scottish Domestic Silverware. *History Scotland Magazine* 2010, 10, (5), pages 27–31.

25. Dietert, R.; Dietert, J., The Impact of Women on the History of Scottish Goldsmithing. *History Scotland Magazine.* 2011, (5), pages 48–53.

26. Dietert, R., Dietert, J., *The "figure-between" goldsmiths of eighteenth-century Edinburgh.* Silver Studies: The Journal of the Silver Society. 2006, 21, pages 49–53.

Chapter 12

1. Feynman, R. P., *Seeking New Laws.* Cornell University Messenger Series Lectures, Lecture 7 In Project Tuva, Microsoft: 1964–65. Accessed November 2, 2012.

2. Palahniuk, C., *Survivor: A Novel.* W.H. Norton & Company: New York, 2010; 289pp page 118.

3. Seneca, L. A., *Ad Lucilium epistulae morales.* With an English translation by Richard M. Gummere (1917) Letter VI: On precepts and exemplars. William Heinemann: London, 1917; page 27 http://archive.org/stream/adluciliume pistu01seneuoft#page/n7/mode/2up. Accessed November 4, 2012.

4. de Bono, E., *Lateral Thinking. Creativity Step by Step.* Harper Colophon: New York, 1973; 304pp.

5. Rickards, T.; Runco, M. A.; Moger, S., *The Routledge Companion to Creativity.* Taylor & Francis US: New York, 2008; 400pp, pages 348–349.

6. Robinson, K., *Out of Our Minds: Learning to Be Creative.* Capstone (Wiley): Chichester, UK 2011; 352pp.

7. de Bono, E., *Serious Creativity: Using the Power of Lateral Thinking to Create New Ideas.* Harper Business: London, 1993; 352pp page 34.

8. Hernandez, J. S.; Varkey, P., Vertical versus lateral thinking. *Physician Exec* 2008, 34, (3), 26–8.

9. Pink, D. H., *A Whole New Mind.* Riverhead Books: New York, 2006; 288pp page 130.

10. Pillai, J. A.; Hall, C. B.; Dickson, D. W.; Buschke, H.; Lipton, R. B.; Verghese, J., Association of crossword puzzle participation with memory decline in persons who develop dementia. *J Int Neuropsychol Soc* 2011, 17, (6), 1006–13.

11. Shah, S.; Lynch, L. M.; Macias-Moriarity, L. Z., Crossword puzzles as a tool to enhance learning about anti-ulcer agents. *Am J Pharm Educ* 2010, 74, (7), 117.

12. Saxena, A.; Nesbitt, R.; Pahwa, P.; Mills, S., Crossword puzzles: active learning in undergraduate pathology and medical education. *Arch Pathol Lab Med* 2009, 133, (9), 1457–62.

13. Manzar, S.; SM, A.-K., Crossword puzzle. A new paradigm for interactive teaching. *Saudi Med J* 2004, 25, (11), 1746–1747.

14. Htwe, T. T.; Sabaridah, I.; Rajyaguru, K. M.; Mazidah, A. M., Pathology crossword competition: an active and easy way of learning pathology in undergraduate medical education. *Singapore Med J* 2012, 53, (2), 121–3.

15. Nombela, C.; Bustillo, P. J.; Castell, P. F.; Sanchez, L.; Medina, V.; Herrero, M. T., Cognitive rehabilitation in Parkinson's disease: evidence from neuroimaging. *Front Neurol* 2011, 2, 82.

16. Newton, P. K.; DeSalvo, S. A., The Shannon entropy of Sudoku matrices. *Proc Roy Soc A* 2010, 466, (2119), 1957–1975.

17. Lucas, R. W., *The Creative Training Idea Book: Inspired Tips and Techniques for Engaging and Effective Learning.* AMACOM Div American Mgmt Assn: New York, 2003; 480pp, pages 124–125.

18. Weisberg, R. W., *Creativity: Understanding Innovation in Problem Solving, Science, Invention, and the Arts.* John Wiley & Sons: Hoboken, NJ, 2006; 640pp, page 20.

19. Medawar, P. B.; Medawar, J.S., *The Life Science.* Wildwood House: London, 1977; 198pp page 138.

20. Tesla, N.; Johnson, B., *My Inventions: The Autobiography of Nikola Tesla.* Hart Brothers: Austin, TX, 1982; 111pp page 61 9780910077002.

21. Dietert, R.; Dietert, J., The Impact of Women on the History of Scottish Goldsmithing. *History Scotland Magazine* 2011, (5), pages 48–53.

22. Dietert, R. R.; DeWitt, J. C.; Germolec, D. R.; Zelikoff, J. T., Breaking patterns of environmentally influenced disease for health risk reduction: immune perspectives. *Environ Health Perspect* 2010, 118, (8), 1091–9.

23. Dietert, R. R., Role of developmental immunotoxicity and immune dysfunction in chronic disease and cancer. *Reprod Toxicol* 2011, 31, (3), 319–26.

Chapter 13

1. Berra, Y., *The Yogi Book: I Really Didn't Say Everything I Said!* Workman Publishing Company: New York, 1998; 127pp page 30.

2. Bobrow, R. S., Evidence for a communal consciousness. *Explore (NY)* 2011, 7, (4), 246–8.

3. Marzouki, Y.; Skandrani-Marzouki, I.; Bejaoui, M.; Hammoudi, H.; Bellaj, T., The contribution of Facebook to the 2011 Tunisian revolution: a cyberpsychological insight. *Cyberpsychol Behav Soc Netw* 2012, 15, (5), 237–44.

4. Farlex Free Dictionary: *synchronicity*. http://www.thefreedictionary.com/synchronicity. Accessed September 3, 2012.

5. Jung, C. G., *Synchronicity: An Acausal Connecting Principle.* Reprint of 1952 edition ed.; Bollingen Foundation: Bollingen, Switzerland, 1993.

6. Cambray, J., *Synchronicity.* Texas A & M University Press: College Station, TX, 2009; 168pp.

7. Koestler, A., *The Roots of Coincidence.* Vintage Books: New York, 1973; 158pp.

8. Hopcke, R. H., *There Are No Accidents. Synchronicity and the Stories of Our Lives.* Riverhead Trade: New York, 1998; 272pp page 122.

9. Williams, G. A., *Demystifying Meaningful Coincidences (Synchronicities): The Evolving Self, the Personal Unconscious, and the Creative Process.* Jason Aronson, Inc: Lanham, MD, 2010; 330pp page 47.

10. Mansfield, V., *Synchronicity, Science, and Soulmaking: Understanding Jungian Syncronicity Through Physics, Buddhism, and Philosphy.* Open Court: Chicago, IL, 1998; 270pp page 24.

11. Maas, J. B.; Robbins, R., *Sleep for Success! Everything You Must Know About Sleep but Are too Tired to Ask.* AuthorHouse: Bloomington, IN, 2011; 264pp.

12. Chopra, D., *The Spontaneous Fulfillment of Desire: Harnessing the Infinite Power of Coincidence.* Harmony Books: New York, 2004; 304pp page 141.

13. Joseph, F.; Graf, D. E., *Synchronicity and You.* Element Books, Inc: Boston, MA, 2003; page ix.

14. Bishop, P., *Synchronicity and Intellectual Intuition in Kant, Swedenborg, and Jung.* Edwin Mellen Press: Lewiston, NY, 2000; 465pp page 361.

15. Scammel, M., *Koestler: The Literary and Political Odyssey of a Twentieth-Century Skeptic.* Random House: New York, 2009; 720pp page 527.

16. McGregor, T.; McGregor, R., The Synchronicity Journal: Your Personal Record of Signs Big and Small. *Adams Media:* 2011; 128pp.

17. Chopra, D., *The Spontaneous Fulfillment of Desire.* Harmony Books: New York, 2004; 304pp.

18. Bartlett, R., *The Physics of Miracles.* Atria Books: New York, 2009; 288pp pages 13–14, 164.

19. Schneiderman, K. *Magical Thinking, Delusions, or Synchronicity?* The Stories We Tell About Mysterious Phenomenon Shape Our Reality Periodical [Online], 2012. http://www.psychologytoday.com/blog/the-novel-perspective/201207/magical-thinking-delusions-or-synchronicity. Accessed September 14, 2012.

Chapter 14

1. Shakespeare, W., *The Tempest*. Arden Shakespeare: London, 1999; 392pp page 254 (Act IV, Scene 1 lines 168–170) http://shakespeare.mit.edu/tempest/full.html. Accessed November 15, 2012.

2. Cooley, M., *City Aphorisms, Fifth Selection*. Pascal Press: New York, 1988; 20pp.

3. Meryman, R., *First Impressions: Andrew Wyeth*. H. N. Abrams: New York, 1991; 92pp page 50.

4. Wieth, M. B.; Zacks, R. T., Time of Day Effects on Problem Solving: When the Non-optimal is Optimal. *Thinking & Resoning* 2011, 17, 387–401.

5. Rodriguiz, T., *Sleepy Brain Think Freely*. Scientific American Mind 2012, 23, (2), 9.

6. Wagner, U.; Gais, S.; Haider, H.; Verleger, R.; Born, J., Sleep inspires insight. *Nature* 2004, 427, (6972), 352–5.

7. Stickgold, R.; Walker, M., To sleep, perchance to gain creative insight? *Trends Cogn Sci* 2004, 8, (5), 191–2.

8. Stampi, C., *Why We Nap. Evolution, Chronobiology, and Functions of Polyphasic and Ultrashort Sleep*. Birkhauser: Boston, 1992; 279 pp.

9. Anthony, W. A., *The Art of Napping*. Larson Publications: Burdett, NY, 1997; 112pp.

10. Caldwell, J. A.; Caldwell, J. L., *Fatigue in Aviation: A Guide to Staying Awake at the Stick*. Ashgate Publishing, Ltd.: Aldershot, UK, 2003; page 124.

11. Mass, J. B.; Wherry, M. L.; Axelrod, D. J.; Hogan, B. R.; Bloomin, J., *Power Sleep: The Revolutionary Program That Prepares Your Mind for Peak Performance*. Villard Books: New York, 1998; 320pp.

12. Maas, J. B.; Robbins, R., *Sleep for Success! Everything You Must Know About Sleep but Are too Tired to Ask*. AuthorHouse: Bloomington, IN, 2011; 264pp.

13. Kelley, T.; Littman, J., *The Ten Faces of Innovation: IDEO's Strategies for Defeating the Devil's Advocate and Driving Creativity Throughout Your Organization*. Doubelday: New York, 2005; 288pp.

14. Gladwell, M., *Blink: The Power of Thinking Without Thinking*. Back Bay Book: New York, 2007; 296pp pages 197–206.

15. Mih, W. C., *The Fascinating Life and Theory of Albert Einstein*. AuthorHouse: Bloomington, IN, 2004; 139pp page ix.

16. Ebert, C.; Ebert., E. S., *The Inventive Mind in Science: Creative Thinking Activities.* Libraries Unlimited: Englewood, CA, 1998; 241pp.

17. O'Neill, J. J., *Prodigal Genius: The Life of Nikola Tesla.* Cosimo: New York, 2007; 336pp page 293.

18. Coren, S., *Sleep Thieves.* Free Press Paperbacks: New York, 1997; 320pp page 284.

19. Martin, P., *Counting Sheep: The Science and Pleasures of Sleep and Dreams.* St. Martin's Press: New York, 2005; 432pp pages 332–333.

20. Gaudreault, A., *American Cinema, 1890–1909: Themes and Variation.* Rutgers University Press: Piscataway, NJ, 2009; 256pp pages 33–34.

21. Horne, J. A., *Sleepfaring: A Journey Through the Science of Sleep.* Oxford University Press: Oxford, 2006; 272pp pages 182–183.

22. Spencer-Lewis, L., *Winston Churchill An American Idol: The X-Factors that Prove the Greatest Briton had Talent.* Kernal & Warden: 2010; 250pp.

23. Hansbury, M., *The Quality of Leadership.* Epitome Books: New Delhi, 2009; 208pp pages 123–124.

24. Schlesinger, A. M., *A Thousand Days: John F. Kennedy in the White House.* First Mariner Books: New York, 2002; 1120pp page 665.

25. Parker, J. J., *Lyndon Johnson.* Xlibris Corporation: Bloomington, IN, 2009; 268pp page 11.

26. Califano, A., *The Triumph & Tragedy of Lyndon Johnson.* PublicAffairs: New York, 2011; 491pp page 28.

27. Beschloss, M. R., *Reaching for Glory: Lyndon Johnson's Secret White House Tapes, 1964–1965.* Touchtone: New York, 2002; 480pp page 22.

28. Reagan, R., *The Reagan Diaries.* HarperCollins: New York, 2009; 784pp pages 13, 16, 345.

29. Gartner, J., *In Search of Bill Clinton: A Psychological Biography.* St. Martin's Press: New York, 2009; 496pp page 145.

30. Restak, R., *Think Smart: A Neuroscientist's Prescription for Improving Your Brain's Performance.* Riverhead Books: New York, 2010; 288pp.

31. Vigeland, C. A., *In Concert: Onstage and Offstage With the Boston Symphony Orchestra.* iUniverse: Lincoln, NB, 2003; 272pp pages 125–129.

32. Cirone, A., *The Great American Symphony Orchestra.* Meredith Music Publications: Galesville, MD, 2011; 208pp.

33. Cowles, F., *The Case of Salvador Dali.* Little Brown: New York, 1960; 334 pp page 66.

34. Caldwell, J. P., *Sleep. Everything You Need to Know.* Firefly Books: Buffalo, NY, 1997; 269pp.

35. Elson, L. C., *Modern Music and Musicians.* The University Society, Inc: New York, 1918; Vol. 8, page 184.

36. Senici, E., *The Cambridge Companion to Rossini.* Cambridge University Press,: Cambridge, 2004; Vol. Part 7, pages 85–103.

37. White, E. W., *Stravinsky the Composer and his Works.* University of California Press: Berkely, CA, 1969; 608pp page 126.

38. Rosselli, J., *The Life of Verdi.* Cambridge University Press: Cambridge, 2000; 220pp page 132.

39. Mednick, S.; Ehrman, M., *Take a Nap! Change Your Life.* Workman Publishing Company: New York, 2006; 140pp pages 114–115.

40. Stickgold, R.; Malia, A.; Maguire, D.; Roddenberry, D.; O'Connor, M., Replaying the game: hypnagogic images in normals and amnesics. *Science* 2000, 290, (5490), 350–3.

41. Ritter, S. M.; Strick, M.; Bos, M. W.; RB, V. A. N. B.; Dijksterhuis, A., Good morning creativity: task reactivation during sleep enhances beneficial effect of sleep on creative performance. *J Sleep Res* 2012, 21, (6), pages 643–7.

42. CDC. *Sleep and Sleep Disorders.* US Government- Centers for Disease Control and Prevention http://www.cdc.gov/sleep/. Accessed November 15, 2012.

43. Dietert, R. R.; DeWitt, J. C.; Germolec, D. R.; Zelikoff, J. T., Breaking patterns of environmentally influenced disease for health risk reduction: immune perspectives. *Environ Health Perspect* 2010, 118, (8), 1091–9.

Chapter 15

1. Feynman, R. P.; Sykes, C. S., *No Ordinary Genius : The Illustrated Richard Feynman.* W. W. Norton & Company: New York, 1995; 272pp, page 161.

2. Budilovsky, J.; Adamson, E., *The Complete Idiot's Guide to Meditation.* 2nd ed.; Alpha Books: Indianapolis, IN, 2002; 384pp.

3. Puddicombe, A., *Get Some Headspace. How Mindfulness Can Change Your Life in Ten Minutes a Day.* St. Martin's Griffin: New York, 2011; 225 pp page 68.

4. NCCAM. *Meditation.* DHHS — National Center for Complementary and Alternative Medicine http://nccam.nih.gov/health/meditation/overview. htm#research. Accessed July 3, 2012.

5. Greenberg, J.; Reiner, K.; Meiran, N., "Mind the trap": mindfulness practice reduces cognitive rigidity. *PLoS One* 2012, 7, (5), e36206.

6. Moore, A.; Malinowski, P., Meditation, mindfulness and cognitive flexibility. *Conscious Cogn* 2009, 18, (1), 176–86.

7. van den Hurk, P. A.; Giommi, F.; Gielen, S. C.; Speckens, A. E.; Barendregt, H. P., Greater efficiency in attentional processing related to mindfulness meditation. *Q J Exp Psychol (Hove)* 2010, 63, (6), 1168–80.

8. Chiesa, A.; Calati, R.; Serretti, A., Does mindfulness training improve cognitive abilities? A systematic review of neuropsychological findings. *Clin Psychol Rev* 2011, 31, (3), 449–64.

9. Tang, Y. Y.; Lu, Q.; Fan, M.; Yang, Y.; Posner, M. I., Mechanisms of white matter changes induced by meditation. *Proc Natl Acad Sci USA* 2012, 109, (26), 10570–4.

10. Moore, A.; Gruber, T.; Derose, J.; Malinowski, P., Regular, brief mindfulness meditationpractice improves electrophysiological markers of attentional control. *Front Hum Neurosci* 2012, 6, 18.

11. Ostafin, B. D.; Kassman, K. T., Stepping out of history: mindfulness improves insight problem solving. *Conscious Cogn* 2012, 21, (2), 1031–6.

12. Brown, C. A.; Jones, A. K., Meditation experience predicts less negative appraisal of pain: electrophysiological evidence for the involvement of anticipatory neural responses. *Pain* 2010, 150, (3), 428–38.

13. Black, D. S.; Cole, S. W.; Irwin, M. R.; Breen, E.; St Cyr, N. M.; Nazarian, N.; Khalsa, D. S.; Lavretsky, H., Yogic meditation reverses NF-kappaB and IRF-related transcriptome dynamics in leukocytes of family dementia caregivers in a randomized controlled trial. *Psychoneuroendocrinology* 2012 doi:10.1016/j.psyneuen.2012.06.11,

14. Dietert, R. R., Misregulated inflammation as an outcome of early-life exposure to endocrine-disrupting chemicals. *Rev Environ Health* 2012, 27, (2–3), 117–31.

15. Zeidan, F.; Martucci, K. T.; Kraft, R. A.; Gordon, N. S.; McHaffie, J. G.; Coghill, R. C., Brain mechanisms supporting the modulation of pain by mindfulness meditation. *J Neurosci* 2011, 31, (14), 5540–8.

16. Nesvold, A.; Fagerland, M. W.; Davanger, S.; Ellingsen, O.; Solberg, E. E.; Holen, A.; Sevre, K.; Atar, D., Increased heart rate variability during nondirective meditation. *Eur J Prev Cardiol* 2012, 19, (4), 773–80.

17. Tsuji, H.; Venditti, F. J., Jr.; Manders, E. S.; Evans, J. C.; Larson, M. G.; Feldman, C. L.; Levy, D., Reduced heart rate variability and mortality risk in an elderly cohort. The Framingham Heart Study. *Circulation* 1994, 90, (2), 878–83.

18. Kemp, A. H.; Quintana, D. S.; Felmingham, K. L.; Matthews, S.; Jelinek, H. F., Depression, comorbid anxiety disorders, and heart rate variability in physically healthy, unmedicated patients: implications for cardiovascular risk. *PLoS One* 2012, 7, (2), e30777.

19. Wegela, K. K., *The courage to be present*. Psychology Today 2010, http://www.psychologytoday.com/blog/the-courage-be-present/201001/how-practice-mindfulness-meditation. Accessed July 25, 2012.

20. Kabat-Zinn, J., *Mindfulness for Beginners. Reclaiming the Present Moment—and Your Life*. Sounds True: Boulder, CO, 2011; 120pp see pages 1, 4, 9, 73.

21. Bair, P. K., *Living from the Heart: Heart Rhythm Meditation for Energy, Clarity, Peace, Joy, and Inner Power.* Crown Publishers: New York, 1998; 336pp.

22. Pearsall, P., *The Heart's Code: Tapping the Wisdom and Power of Our Heart Energy.* Random House Digital, Inc: New York, 1998; 288pp.

23. Thurman, H., *Meditations of the Heart*. Beacon Press: Boston, 1999; 216pp.

24. Bartlett, R., *The Physics of Miracles*. Atria Books: New York, 2009; 288pp pages 13–14, 164.

25. Vessantara, *The Heart: The Art of Meditation*. Windhorse Publications: Cambridge, UK, 2006; 164pp.

26. Burleson, K. O.; Schwartz, G. E., Cardiac torsion and electromagnetic fields: the cardiac bioinformation hypothesis. *Med Hypotheses* 2005, 64, (6), 1109–16.

27. Manev, H., The heart-brain connection begets cardiovascular psychiatry and neurology. Cardiovasc Psychiatry Neurol 2009, 546737.

28. Bradley, R. T.; Atkinson, M.; Rees, R. A.; Tomasino, D.; Galvin, P. *Facilitating Emotional Self-Regulation in Preschool Children: Efficacy of the Early HeartSmarts Program in Promoting Social, Emotional and Cognitive Development;* HeartMath Research Center, Institute of HeartMath: Boulder City, CA, 2009; 36pp. Research, http://www.heartmath.org/templates/ihm/downloads/pdf/research/publications/facilitating-emotional-self-regulation-in-preschool-children.pdf. Accessed September 20, 2012.

29. Seaward, B. L., *Quiet Mind, Fearless Heart: The Taoist Path through Stress and Spirituality.* John Wiley & Sons: Hoboken, NJ, 2004; pages 83–84.

30. Vas, L. S. R., *Meditation*. Pustak Mahal: Delhi, 2001; 224pp pages 169–174.

31. Shapiro, D. H. J.; Walsh, R. N., Meditation: Classic and Contemporary Perspectives. *Transaction Publications: Piscataway*, NJ, 2008; 744pp pages 376–377.

32. Bromley, D. G.; Cowan, D. E., *Cults and New Religions: A Brief History.* Wiley-Blackwell: Malden, MA, 2008; 272pp.

33. Rosenthal, N. E., *Transcendence: Healing and Transformation Through Transcendental Meditation.* Penguin Books: New York, 2011; 303pp.

34. Anderson, J. W.; Liu, C.; Kryscio, R. J., Blood pressure response to transcendental meditation: a meta-analysis. *Am J Hypertens* 2008, 21, (3), 310–6.

35. Goldstein, C. M.; Josephson, R.; Xie, S.; Hughes, J. W., Current perspectives on the use of meditation to reduce blood pressure. *Int J Hypertens* 2012, 2012, 578397.

36. Jayadevappa, R.; Johnson, J. C.; Bloom, B. S.; Nidich, S.; Desai, S.; Chhatre, S.; Raziano, D. B.; Schneider, R., Effectiveness of transcendental meditation on functional capacity and quality of life of African Americans with congestive heart failure: a randomized control study. *Ethn Dis* 2007, 17, (1), 72–7.

37. Walton, K. G.; Fields, J. Z.; Levitsky, D. K.; Harris, D. A.; Pugh, N. D.; Schneider, R. H., Lowering cortisol and CVD risk in postmenopausal women: a pilot study using the Transcendental Meditation program. *Ann N Y Acad Sci* 2004, 1032, 211–5.

Chapter 16

1. Sagan, C., *Blues for a Red Planet.* In Cosmos, USA, 1980; Episode 5, 60 minutes.

2. Stevens, T. G., *You Can Choose to be Happy in Life: Rise Above Anxiety, Anger, and Depression.* Wheeler-Sutton Publishing Co: Palm Desert, CA, 2010; 320pp pages 52–53.

3. Lightman, D., *Power Optimism: Enjoy the Life You Have, Create the Success You Want.* Power Optimism LLC: Abington, PA, 2004; 212pp pages 121–122.

4. Rieger, A., *Soap Operas.* GRIN Verlag: Munich, 2002; 14pp page 2.

5. Dietert, R. R., Developmental immunotoxicology: focus on health risks. *Chem Res Toxicol* 2009, 22, (1), 17–23.

6. Dietert, R. R.; DeWitt, J. C.; Germolec, D. R.; Zelikoff, J. T., Breaking patterns of environmentally influenced disease for health risk reduction: immune perspectives. *Environ Health Perspect* 2010, 118, (8), 1091–9.

7. Dietert, R. R.; Dewitt, J. C.; Luebke, R. W., *Reducing the prevalence of chronic diseases.* In Immunotoxicity, Immune Dysfunction and Chronic Disease, Dietert, R. R.; Luebke, R. W., Eds. Springer: New York, 2012; pages 419–440,

8. Dietert, R. R., Fractal immunology and immune patterning: potential tools for immune protection and optimization. *J Immunotoxicol* 2011, 8, (2), 101–10.

9. Murphy, J., *Inner Excellence: Achieve Extraordinary Business Success Through Mental Toughness.* McGraw Hill Professional: New York, 2009; 272pp pages 88, 125, 145, 245.

Chapter 17

1. Maurois, A., *The Life of Sir Alexander Fleming: Discoverer of Penicillin.* Pengiun Books: New York, 1963; 328pp pages 120, 234.

2. LeFevre, D. N., *Best New Games.* Human Kinetics: Champaign, IL, 2002; 217pp pages 8–9.

3. McKenna, K. *Interview with David Lynch taken from Colección Imagen: conducted March 8, 1992.* Periodical [Online], 1992. http://www.thecity-fabsurdity.com/intpaint.html. Accessed August 12, 2012.

4. Ackerman, D., *Deep Play.* Vintage: New York, 2000; 256pp page 6.

5. Doorley, S.; Scott Witthoft, S., *Make Space: How to Set the Stage for Creative Collaboration.* Wiley: Hoboken, NJ, 2011; 272pp.

6. Capps, D., Child's play: The creativity of older adults. *J Relig Health* 2012, 51, (3), 630–50.

7. Mammas, I. N.; Spandidos, D. A., A 3,000-year-old child's toy. *Eur J Pediatr* 2012, 171, (9), 1413.

8. Beck, M., *Finding Your Way in a Wild New World.* Simon & Schuster: New York, 2011; 320pp pages 26–29.

9. Stone, L. *Finding Oursleves Though Play.* Periodical [Online], 2010. http://lindastone.net/2010/01/03/finding-ourselves-through-play/. Accessed November 6, 2012.

10. Dunn, R. *Painting With penicillin: Alexander Fleming's germ art*. Periodical [Online], 2010. http://www.smithsonianmag.com/science-nature/Painting-With-Penicillin-Alexander-Flemings-Germ-Art.html. Accessed October 16, 2012.

11. Root-Bernstein, R.; Root-Bernstein, M., *Sparks of Genius*. Mariner Books: Boston, 2001; 416pp pages 246–248.

12. Eger, J., *Einstein's Violin*. Jeremy P Tarcher/Pengiun: New York, 2005; 432pp.

13. Foster, B., *Einstein and His Love of Music*. Physics World 2005, page 34 http://www.pha.jhu.edu/einstein/stuff/einstein&music.pdf. Accessed October 15, 2012.

14. Flotteron, N. *Sailor And Scientist: Albert Einstein Makes History On The East End*. Periodical [Online], 2010. http://www.hamptons.com/Home-And-Garden/East-End-Heirlooms/12852/Sailor-And-Scientist-Albert-Einstein-Makes.html#. UH21rGfheJ4. Accessed October 16, 2012.

15. The Da Vinci Machines Exhibition, Adelaide Showgrounds, *The Da Vinci Machine Exhibition: Brenton Ragless is amazed by the creations of this Italian artist, scientist and inventor in the Adelaide City region of South Australia*. In Postcards Online Channel 9 Adelaide, Australia: 2012; http://www.postcards-sa.com.au/features2011/da_vinci.html October 24, 2012.

16. The Da Vinci Machines Exhibition, *Denver Pavillions, Discover the Da Vinci in You*. In *2012*; http://www.davinciexhibitdenver.com/. Accessed October 24, 2012.

17. Museum of Leonardo Da Vinci. Florence Italy. Homepage and Exhibitions. http://translate.google.com/translate?hl=en&sl=it&u=http://www.mostredi-leonardo.com/&prev=/search%3Fq%3Dmuseum%2Bof%2Bleonardo%2Bda%2Bvinci%2Bflorence%26hl%3Den%26biw%3D1366%26bih%3D599%26prmd%3Dimvnso&sa=X&ei=F3WIUIuSKcq20QGjhIGACw&ved=0CC0Q7gEwAA. Accessed October 24, 2012.

18. Brizio, A. M., Notebooks. *New aspects of Leonardo's genius emerge from his rediscovered manuscripts*. The UNESCO Courier October 1974, 1974, 8–10 http://unesdoc.unesco.org/images/0007/000748/074877eo.pdf. Accessed October 12, 2012.

19. Feynman, R. P.; Leighton, R.; Hutchings, E., *Surely You're Joking, Mr. Feynman!* W.W. Norton & Company: New York, 1985; 350pp pages 137–155.

20. Royal Swedish Academy of Sciences. *Press Release. Graphene — the perfect atomic lattice.*Periodical [Online], 2010. http://www.nobelprize.org/nobel_prizes/physics/laureates/2010/press.html. Accessed October 17, 2012.

21. Hagedorn, R. K., *Benjamin Franklin and Chess in Early America: A Review of the Literature.* University of Pennsylvania Press: Philadelphia, 1958; 92pp.

22. Franklin, B., *On The Morals of Chess.* The Columbian Magazine 1786, see page 1 http://meritbadge.org/wiki/images/6/65/Ben_Franklin_-_Morals_of_Chess.pdf. Accessed November 30, 2012.

23. Pasles, P. C., The Lost Squares of Dr. Franklin: Ben Franklin's Missing Squares and the Secret of the Magic Circle. *American Mathematical Monthly* 2001, 108, (6), 498–511.

24. Witkin, E., Enelyn Witkin on Babara McClintock Sense of Humor. In Oral History Collection, Cold Spring Harbor National Laboratories: 2000; http://library.cshl.edu/oralhistory/interview/cshl/barbara-mcclintock/barbara-mclintock-personality/. Accessed November 15, 2012.

25. Peterson, T., *A Celebration of the Life of Dr. Barbara McClintock: Renowned Geneticist Remembered at Memorial Service.* Probe 1993, 3, 1–2.

26. Witkin, E., Evelyn Witkin on Barbara McClintock Personality. In Oral History Collection, Cold Spring Harbor Laboratories: 2000; http://library.cshl.edu/oralhistory/interview/cshl/barbara-mcclintock/barbara-mclintock-personality/. Accessed November 15, 2012.

Chapter 18

1. Viereck, G. S., What Life Means to Einstein. *Saturday Evening Post* 1929, page 113.

2. Tukey, J. W., *Exploratory Data Analysis.* Addison-Wesley Publishing Company: Boston, 1977; 688pp.

3. Andreasen, N. C.; Ramchandran, K., Creativity in art and science: are there two cultures? *Dialogues Clin Neurosci* 2012, 14, (1), 49–54.

4. Robinson, K., *Out of Our Minds: Learning to Be Creative.* Capstone (Wiley): Chichester, UK 2011; 352pp, page 158.

5. Ness, R. B., *Innovation Generation.* Oxford University Press: Oxford, 2012; 272pp.

6. Jenny, H., *Cymatics. A Study of Wave Phenomena and Vibration.* Macromedia: State College, PA, 2001; 295pp.

7. Mannes, E., *The Power of Music: Pioneering Discoveries in the New Science of Song.* Walker Publishing Company: New York, 2011; 288pp.

8. Leeds, J., *The Power of Sound: How to Be Healthy and Productive Using Music and Sound.* Healing Arts Press: Rochester, VT, 2010; 302pp.

9. Morris, W., *The Dream of Color Music, And Machines That Made it Possible.* Animation World Magazine 1997, http://www.awn.com/mag/issue2.1/articles/moritz2.1.html. Accessed November 7, 2012.

10. Laszlo, A., *Die Farblichtmusik.* Breitkopf & Hartel: Leipzig, 1925; 71pp.

11. Helfgott, J.; Middleton, N., *Music and the Brain: Halt or I'll Play Vivaldi.* In Library of Congress: 2009; http://www.loc.gov/podcasts/musicandthebrain/podcast_jacquelinehelfgott.html. Accessed October 9, 2012.

12. Gvaryahu, G.; Cunningham, D. L.; Van Tienhoven, A., Filial imprinting, environmental enrichment, and music application: effects on behavior and performance of meat strain chicks. *Poultry Science* 1989, 68, (2), 211–217.

13. Davila, S. G.; Campo, J. L.; Gil, M. G.; Prieto, M. T.; Torres, O., Effects of auditory and physical enrichment on 3 measurements of fear and stress (tonic immobility duration, heterophil to lymphocyte ratio, and fluctuating asymmetry) in several breeds of layer chicks. *Poult Sci* 2011, 90, (11), 2459–66.

14. Itzkoff, D., *Hans Zimmer Extracts the Secrets of the 'Inception' Score.* New York Times July 28, 2010, 2010, http://artsbeat.blogs.nytimes.com/2010/07/28/hans-zimmer-extracts-the-secrets-of-the-inception-score/. Accessed November 7, 2012.

15. Palisca, C. V. *Vincenzo Galilei.* Periodical [Online], 2007. http://www.oxfordmusiconline.com/subscriber/article/grove/music/10526?q=Vincenzo+Galilei&search=quick&pos=1&_start=1#firsthit. Accessed November 16, 2012.

16. Valleriani, M., *Galileo Engineer.* Springer: New York, 2010; 320pp pages 12–13.

17. Foster, B., *Einstein and His Love of Music.* Physics World 2005, page 34 http://www.pha.jhu.edu/einstein/stuff/einstein&music.pdf. Accessed October 15, 2012.

18. Isaacson, W., *Einstein: His Life and Universe.* Simon & Schuster: New York, 2007; 704pp page 14.

19. Einstein, A., *The Ultimate Quotable Einstein.* Princeton University Press: Princeton, NJ, 2011; 576pp pages 240–241 edited by Alice Calaprise.

20. Winternitz, E., *Leonardo Da Vinci as a Musician.* Yale University Press: New Haven, CT, 1982; 241pp.

21. Wright, E.; Franklin, B., *Benjamin Franklin: His Life As He Wrote It.* Harvard University Press: Cambridge, MA, 1990; 297pp pages 157–161.

22. Grenander, M. E., Reflections on the String Quartet(s) Attributed to Franklin. *American Quarterly* 1975, 27, (1), 73–87.

23. Teller, E.; Schoolery, J., *Memoirs: A Twentieth Century Journey In Science And Politics.* Perseus Books: Cambridge, MA, 2002; 672pp page 168.

24. Schiller, J., *NanoTechnology Development.* CreateSpace Independent Publishing Platform: Seattle, WA, 2010; 250pp pages 27–28.

25. Mehra, J., *The Beat of a Different Drum: the Life and Science of Richard Feynman.* Oxford University Press: Oxford, 1996; 710pp.

26. Gribbin, J.; Gribbin, M., *Richard Feynman: A Life in Science.* Viking: New York, 1997; 320pp page 208.

27. Feynman, R. P.; Leighton, R.; Hutchings, E., *Surely You're Joking, Mr. Feynman! : adventures of a curious character.* W.W. Norton: New York, 1985; 350pp see pages 170–171 and 323–329.

28. Feynman, R. P., *The Law of Gravitation — An Example of Physical Law.* In Richard Feynman — Cornell University Messenger Series, BBC: 1964; http://research.microsoft.com/apps/tools/tuva/index.HTML#data=4%7C6b89dded-3eb8-4fa4-bbcd-7c69fe78ed0c%7C%7C. Accessed November 6, 2012.

29. Siegel-Itzkovich, J., *Of Mice, Molecules and Music.* Jerusalem Post April 23, 2012, http://www.jpost.com/HealthAndSci-Tech/ScienceAndEnvironment/Article.aspx?id=90817. Accessed November 7, 2012.

30. Cassutt, M., *Max Q Live. In Space No One Can Hear You Sing.* Smithsonian Museums Air and Space Magazine 2009, see page 5 http://www.airspacemag.com/space-exploration/Max-Q-Live.html. Accessed October 15, 2012.

31. Robinson, K.; Aronica, L., *The Element: How Finding Your Passion Changes Everything.* Viking: New York, 2009; 272pp.

32. Root-Bernstein, M.; Root Berstein, R. *Dance Your Experiment.* Periodical [Online], 2012. http://www.creativitypost.com/science/dance_your_experiment. Accessed October 19, 2012.

33. Bohannon, J., *Dance Your Ph.D. Winner Announced.* Science Magazine 2011, http://news.sciencemag.org/sciencenow/2011/10/dance-your-phd-winner-announced.html. Accessed November 11, 2012.

34. Bohannon, J., *Why do scientists dance?* Science 2010, 330, (6005), 752 http://www.ncbi.nlm.nih.gov/entrez/query.fcgi?cmd=Retrieve&db=PubMed&dopt=Citation&list_uids=21051611. Accessed November 11, 2012.

35. Berg, P., *Dance of the Ribosomes*. In Stanford University Department of Chemistry: 1971; http://cmgm.stanford.edu/movie/. Accessed Novermber 16, 2012.

36. Madrigal, A., *Dance as Algorithm: What Happens When an Animated GIF Comes to Life*. The Atlantic July 26, 2012, 2012, http://www.theatlantic.com/technology/archive/2012/07/dance-as-algorithm-what-happens-when-an-animated-gif-springs-to-life/260336/. Accessed November 6, 2012.

37. Fink, A.; Graif, B.; Neubauer, A. C., Brain correlates underlying creative thinking: EEG alpha activity in professional vs. novice dancers. *Neuroimage* 2009, 46, (3), 854–62.

38. Noy, L.; Dekel, E.; Alon, U., The mirror game as a paradigm for studying the dynamics of two people improvising motion together. *Proc Natl Acad Sci USA* 2011, 108, (52), 20947–52.

39. Powell, K., *Science at the Improv*. In Howard Hughes Medical Institute Bulletin [Online] 2007. http://www.hhmi.org/bulletin/may2007/chronicle/improv.html. Accessed November 7, 2012.

40. Koestler, A., *The Act of Creation*. Penguin/Arkana: New York, 1990; 752pp.

41. Careri, G., Physicists and Painters: the Similar Search for Meaning. *Leonardo* 1989, 22, 113–115.

42. Heck, A., *Creativity in Arts and Sciences: A Survey*. In: Organizations and Strategies in Astronomy II Heck, A., Ed. Kluwer Academic Publishers: Dordrecht, The Netherlands, 2001; pages 257–268,

43. Leibowitz, J. R., Hidden Harmony: *The Connected Worlds of Physics and Art*. John Hopkins University Press: Baltimore, MD, 2008; 160pp.

44. Shier, A., *Science Museum Roundup*. Chicago Art Magazine July 10, 2010, 2010, http://chicagoartmagazine.com/2010/07/science-museum-roundup/. Accessed October 21, 2012.

45. Karlic, H.; Baurek, P., Epigenetics and the power of art. *Clin Epigenetics* 2011, 2, (2), 279–82.

46. Neumann, R., *Bedside Arts: The Role of Humanities in Hospital Care*. The Hospitalist 2006, http://www.the-hospitalist.org/details/article/239589/Bedside_Arts.html. Accessed November 7, 2012.

47. Sullivan, A. M., Art at the Mayo Clinic. *Mayo Clinic Proceedings* 2001, 76, (2), 153.

48. Shier, A., *International Museum of Surgical Science*. Chicago Art Magazine July 28, 2010, http://chicagoartmagazine.com/2010/07/international-museum-of-surgical-science/. Accessed October 21, 2012.

49. Shier, A., *Scientists as Artists*. Chicago Art Magazine August 6, 2010, http://chicagoartmagazine.com/2010/08/scientists-as-artists/. Accessed October 21, 2012.

50. Mitchell, K. *The Fractal Art Manifesto*. Damien M. Jones https://www.fractalus.com/info/manifesto.htm.Accessed April 22, 2012.

51. Humphreys, R.; Eagan-Deprez, K., Fostering mind-body synchronization and trance using fractal video. *Technoic Arts* 2005, 3, (2), 93–105.

52. Draves, S., Art feature for 2008: electric sheep. *Nonlinear Dynamics Psychol Life Sci* 2008, 12, (1), 131.

53. Hoffmann, R., *The Metamict State*. University Presses of Florida, University of Central Florida Press: Orlando, 1987; 104pp.

54. Hoffmann, R., *Gaps and Verges*. University Presses of Florida, University of Central Florida Press: Orlando, 1990; 89pp.

55. Hoffmann, R., *Memory Effects*. Calhoun Press, Columbia College, Dept. of Art and Design: Chicago, 1999; 80pp.

56. Djerassi, C.; Hoffmann, R., *Oxygen: a play in two acts*. Wiley-VCH: Weineim, Germany, 2001; 119pp.

57. Cornelia Street Café, *Entertaining Science*. Cornelia Street Café. New York Cityhttp://corneliastreetcafe.wordpress.com/entertaining-science/. Accessed November 16, 2012.

Chapter 19

1. Moore, R., *Niels Bohr: The Man, His Science, & the World They Changed*. Alfred A. Knopf: New York, 1966; 436pp page 196.

2. Maxwell, J. C., *The Scientific Papers of James Clerk Maxwell. Volumes 1 & 2*. Reprint ed.; Dover Publications: 1952; page 243 from Lecture on Introductory Physics.

3. Pais, A., *The Genius of Science: A Portrait Gallery*. Oxford University Press: Oxford, 2000; 356pp page 24.

Index